GRAINING
ANCIENT AND MODERN

Yours Respectfully,
William E. Wall

GRAINING

ANCIENT AND MODERN

by

WILLIAM E. WALL

Preface by J. M. Ratcliffe

SHERIDAN HOUSE

This edition first published in the United States
of America 1988 by Sheridan House Inc.,
145 Palisade St., Dobbs Ferry, N.Y. 10522

All rights reserved. No part of this publication
may be reproduced or transmitted, in any form or
by any means, without permission.

First published 1905

ISBN 0 911378 79 0

Printed in England

PREFACE TO THIS EDITION

In recent years there has been a revival of interest in special decorative effects, such as graining, in the construction and wood-working industries, and a number of books has been republished on the subject. This rare work, however, stands pre-eminent as a source of special information about graining, and is now made available again in a modern edition, complete with colour plates, for the first time in many decades.

This reprint of the second edition has not been revised in any substantive way, thereby correcting any defects of subsequent editions of the book up to the late 1940s which had sought to re-interpret the information and, in consequence, omitted valuable original references to source material, and also work-histories of contemporary craftsmen.

Although it first appeared more than fifty years ago, this edition contains basic material of both practical and reference use. The techniques, tools and brushes described are still valid for use today, and provide foundation knowledge for emulating the finishes of the Victorian era, (though I would caution against using the formula for megilp quoted from Nathaniel Whittock's *Painter's and Glazier's Guide*, despite the fact that Whittock states that it was used in the preparation of the illustrations in his book!).

My own special interest in seeing the book made available once again stems from my company's 77 years of manufacturing proprietary scumbles and glazing materials for the craftsmen, and, in more recent years, ensuring the continued availability of the specialised tools and brushes required. With this book and the related materials, the techniques refined by the author from many years' practical experience in the nineteenth century can now be reproduced more widely today.

William Wall had an extensive knowledge of his subject and employed his skill to great effect in a variety of media. I hope that this book will help readers to learn new techniques and to build on the skills they already possess.

Jeffrey M Ratcliffe
1988

PREFACE

The chief object of the writer of this book is to provide instruction for those of our trade, especially young men, who desire to become proficient in graining, for in these days little opportunity is provided to learn any trade in the building line, least of all in the specialty of graining.

A series of articles by the writer entitled "Practical Graining" were published in 1889–1890 in the *House Painting and Decorating Magazine* of Philadelphia, Pennsylvania. They were later issued in book form and have been out of print for several years.

Frequent requests for copies have induced the writer to issue the present volume, which will be found to contain added information on this subject.

My thanks are due to Mr. James Kay of St. Louis, Missouri, for information concerning the antiquity of graining; also to The W. J. Dobinson Engraving Co. of Boston, for excellent plates; and to The Norwood Press for excellence of typography and composition.

WILLIAM E. WALL

1905

CONTENTS

CHAPTER		PAGE
I.	Antiquity of Graining.	1
II.	Imitations	4
III.	Eminent Grainers of the Last Century	7
IV.	Ground-colors	11
V.	Graining Colors	16
VI.	Thinners for Oil Colors	17
VII.	Megilp for Oil Color	18
VIII.	Tools	20
IX.	Rubbing in Oil Color	21
	Rubbing in Water Color	24
X.	Combing in Oil Color	25
	Combing in Water Color	29
XI.	Graining Crayons	31
XII.	Wiping out Heart Grains in Oil Color	32
XIII.	Curly Maple	33
XIV.	Bird's-eye Maple	35
XV.	Silver Maple	40
XVI.	Whitewood	41
XVII.	Satinwood	42
XVIII.	White Mahogany	45
XIX.	Ash	46
	Hungarian Ash	51
	Burl Ash	53
XX.	Quartered Oak	54
	Wiping out in Oil Color	57
	Graining Quartered Oak	58
	Fourteen Ways of imitating Quartered Oak	60

ix

CONTENTS

CHAPTER		PAGE
XXI.	ENGLISH OAK	65
	POLLARD OAK	66
	ROOT OF OAK	67
XXII.	HEART, OR SAP, OAK	67
	WIPING OUT HEART GRAINS OF OAK	69
XXIII.	CHESTNUT	71
XXIV.	WHITE OREGON CEDAR	72
XXV.	YELLOW PINE	74
	PITCH PINE, OR HARD PINE	75
XXVI.	CYPRESS	77
XXVII.	QUARTERED SYCAMORE	78
XXVIII.	CHERRY	81
XXIX.	CURLY BIRCH	85
XXX.	BLACK WALNUT	87
	CRAYONS FOR WALNUT GRAINING	90
	CURLY WALNUT	91
	FRENCH WALNUT BURL	91
	ITALIAN WALNUT	93
	CIRCASSIAN WALNUT	93
XXXI.	MAHOGANY	94
XXXII.	TEAK	98
XXXIII.	ROSEWOOD	99
XXXIV.	OVERGRAINING	101
XXXV.	CEILINGS	102
XXXVI.	FLOORS	104
	MANILA PAPER FOR COVERING A POOR FLOOR	106
	VARNISHING A GRAINED FLOOR	108
XXXVII.	PATENT GRAINING DEVICES	108
XXXVIII.	SHOW PANELS	111
XXXIX.	GRAINING ON GLASS	113
XL.	IMITATIONS OF CARVING	114
	IMITATIONS OF MOULDINGS	114
XLI.	CAUSES OF CRACKING IN GRAINED WORK	116

CONTENTS

CHAPTER		PAGE
XLII.	The Grainer in Fiction	118
XLIII.	Graining a Door Quartered Oak	121
XLIV.	New Methods	124
XLV.	Journeymen	127
XLVI.	Bicycle for City or Country Work	129
XLVII.	Butternut	129
Index		133

PLATES

In this edition the plates are grouped in a single sequence, and – as was the case in the first edition – the sequence in which they appear departs in a few instances from strict numerical order, so as to relate more closely to references in the text and the grouping of the wood types.

PLATE

Author's Photograph, with Signature. *Frontispiece.*
1. Tools:
    Cut (Photographic):
        1, stippler; 2, large overgrainer; 3, short-haired overgrainer; 4, bone comb (for separating bristles of overgrainer); 5, large mottler; 6, small mottler; 7, badger blender; 8, piped overgrainer.
2. Cut:
    1, flat fresco bristle liner; 2, short-haired liner or fitch tool; 3, sash tool (oval); 4, rubbing-in brush; 5, check roller; 6. sponge; 7, fine steel split comb; 8, medium steel comb; 9, medium split steel comb; 10, coarse steel comb; 11, selvedge of straw matting; 12, rubber comb.
3. Large Oval Table-top (Compass Centre).
4. Smaller Oval Table-top (Card Centre).
5. "Jersey Oak," Cambridge, Mass., done in 1845.

PLATE
6. Curly Maple — Mottled to Overgrain.
7. Curly Maple — Mottled and Overgrained.
8. Bird's-eye Maple — First Stage.
9. Bird's-eye Maple — Overgrained.
10. Bird's-eye Maple — Finished.
11. Whitewood.
12. Satinwood — Feather Panel.
13. White Mahogany.
14. Quartered Sycamore.
15. Light Ash — Wiped out and
15 A.   Pencilled.
16. Ash — Hungarian Ash Panel.
17. Ash — Burl Panel.
18. Dark Ash.
19. Chestnut.
20. How Quartered Oak is Sawed.
21. Oak — Combed, ready for Quartered Veins.
22. Light Quartered Oak — Overgrained.
23. Light Quartered Oak — Overgrained.
24. Quartered Oak — Dark Panel.
25. Quartered Oak.
26. Quartered Oak — in Water Color.

PLATES

PLATE
- 27. Quartered Oak — Light.
- 28. Light Quartered Oak.
- 29. English Quartered Oak — Pollard Oak Panel.
- 29 A. English Oak — Root of Oak Panel.
- 30. Dark Quartered Oak.
- 31. Dark Quartered Oak.
- 32. Dark Quartered Oak.
- 33. Heart of Oak — Light.
- 34. Dark Heart of Oak.
- 35. Cherry — Mottled; ready to Overgrain.
- 36. Cherry — Mottled and Overgrained.
- 37. Cherry — Mottled and Overgrained.
- 38. Butternut.
- 39. Cypress.
- 40. Curly Birch.

PLATE
- 41. Oregon Cedar.
- 42. Yellow Pine.
- 43. Pitch Pine, or Hard Pine.
- 44. Mahogany — Mottled Panel.
- 45. Mahogany — Figured.
- 46. Mahogany — Feathered Panel.
- 46 A. Mahogany — Feathered Panel.
- 47. Teak.
- 48. Stippling for Walnut or Mahogany.
- 49. Black Walnut — Overgrained.
- 50. Curly Walnut.
- 51. Black Walnut — Burl Panel.
- 52. Italian Walnut.
- 53. Circassian Walnut.
- 54. Rosewood — First Stage.
- 55. Rosewood — Overgrained.
- 56. Imitation of Carving.

GRAINING, ANCIENT AND MODERN

CHAPTER I

ANTIQUITY OF GRAINING

RECENT discoveries in Egypt and elsewhere show the inherent disposition of the earliest races of mankind to provide themselves with imitations of precious stones, etc., but it is not generally known that the ancient Egyptians were expert grainers.

In a book entitled "Museum of Antiquity," a description of ancient life three thousand years ago written by L. W. Yaggy, M.S., and T. L. Haines, A.M., and published by Weaver and Company, in Kansas City, Missouri, and Chicago, Illinois, in 1882, we read on page 350, — "Carpenters and cabinet makers were a numerous class of workmen and their occupation forms one of the most important subjects in the paintings which represent the Egyptian trades." "For ornamental purposes, even sometimes for doors and boxes, foreign woods were employed. Deal and cedar were imported from Syria, and part of the contributions exacted from the conquered tribes of Ethiopia and Asia consisted in ebony and other rare woods which were annually brought by the chiefs deputed to present their countries' tribute to the Egyptian Pharaohs." "Boxes, chairs, tables, sofas, etc., were often made of ebony inlaid with ivory, sycamore, and acacia veneering, with thin layers and carved devices of rare wood added as ornament on inferior surfaces; and a fondness for display induced the Egyptians to paint common boards to imitate foreign varieties, so generally practised in other countries at the present day. The colors were usually applied on a thin coating of stucco or a ground smoothly laid on prepared wood, and the various knots and grains made to resemble the wood they intended to counterfeit."

This account would appear to indicate that grainers were a professional class of artisans over three thousand years ago.

There is shown in the British Museum in London a bill of account several centuries old, for painting and graining a room in the Tower of London.

Undoubtedly, the art of graining came to England from Continental Europe several centuries ago, and, without doubt, its finest exemplars have been developed in the "tight little isle." Their progeny have gone all over the world, and wherever they have travelled have "made their mark."

The banishment of the Huguenots from France (A.D. 1666–1789) and their settlement in England has been referred to by some writers as a period when England's skilled workmen received a great stimulus, and many of the arts and crafts were benefited by recruits added to their ranks from these skilled artisans of France. Doubtless there were among them skilful grainers who imparted to the trade the impress of their skill; but I feel sure the beginning of graining in England does not date from the days of the exodus of the Huguenots from France, but that it had existed long before their advent.

Examples of imitations of wood and marble are said to be found in the old cities of Continental Europe, which date back for several centuries. As a rule, nearly all the ancient graining was done in distemper or water colors, and when properly protected, it is in all respects as durable as oil colors, and many woods can be more successfully imitated in water colors than in oil colors.

One of the old school of water-color artists of the

eighteenth century, whose name I do not now recall, was a great opponent of the school of oil painters. He claimed that it was folly to paint pictures in oil colors, as the seeds of destruction were sown in every brush of oil color which was applied to canvas and in time it would certainly crack and destroy itself; while water-color work, which contained no gums or oil varnishes, would remain durable for centuries and its colors would remain comparatively unimpaired.

I have seen water-color graining done sixty-two years with but little varnish applied over it (which doubtless was its chief salvation) which was in good condition at the end of that time.

The Chinese and Japanese were probably among the earliest imitators of wood. I have seen a cheap cabinet bought in China, which appeared to be made of fine-grained cedar, but which, on careful examination, proved to be made of inferior wood covered with very thin rice paper on which had been printed an excellent imitation of figured cedar-wood. The cabinet was finished without varnish, and the effect was similar to that of smoothly finished cedar-wood.

It would be safe to say that such imitations had been done by hand or printed for many centuries in both China and Japan.

House painters were among the earliest artisans who came to America, and some of them must have possessed a knowledge of graining as it was practised in the larger cities of Great Britain or Continental Europe; their descendants have carried on the work in the larger cities of the country, and their numbers have been increased from the constant stream of emigrants who have been arriving each year.

OLD-FASHIONED GRAINING

The panel on the opposite page (Plate 5) is a photographic copy of work done in the first High School building,

Cambridge, Massachusetts, in 1845. It was called "Jersey Oak." This work was done in water colors and is a type of the average job of that time. Later, in the fifties, the advent of several British grainers to Boston, who were expert workmen in both oil and water color, showed the possibilities of good work and gave the trade a high standard of workmanship which has not been wholly lost to this day, although all but one of these men have passed to the great majority. Between 1850 and 1860 there came to Boston, Massachusetts, Walter and William J. McPherson of Scotland, William Munro Ross of Scotland, William Hopson of England, and James Keleher of Ireland. These men all followed graining for the trade, and all were expert workmen.

CHAPTER II

IMITATIONS

NICCOLO POUSSIN, the eminent painter, born at Audsly, Normandy, in 1594, declares that "Painting is an imitation, by means of lines and color on some superfices, of anything that can be seen under the sun, its end to please, — principles that every man capable of reasoning may learn."

Here is a plain, unvarnished statement from an eminent authority directly bearing on the question of imitation, and to the ordinary mortal it would appear to be a sensible one. If it be true, as the poet Keats wrote, "A thing of beauty is a joy forever" (a statement that few will antagonize), why condemn a cleverly executed imitation of wood or marble, which admittedly is beautiful, on the ground that it is "false in art," when the best of painted art, according to the above authority, is merely an imitation.

It has often been affirmed that all art is, in a sense, imitative, and it requires an extremely fine perception to discover the dividing line separating the talent required to execute a finely grained panel from that required to paint a picture.

Despite all that has been said and written against graining, on the ground that it should not be tolerated, being a sham and a deception and thus reprehensible, it will always find its place in decorative work, and when properly done will plead its own cause.

I have noticed in the last twenty years a growing tendency on the part of some distinguished Royal Academicians to make a background of imitation marble one of the features of their pictures, especially on large canvases. Is it any less artistic to paint in imitation of marble on the wall of a staircase than on the canvas of a picture?

We are told that the fault of the painted marble on the wall is that its primary intent is to deceive. Admitting that, is it any less beautiful? And must we be content with plain surfaces of unadorned paint for fear that if we grain or marble them, some one may be deceived thereby?

Suppose a lady buys a set of oak furniture for her dining room. The table, chairs, and sideboard are made of quartered oak; now if the woodwork of that room has been painted white or stained red, or whatever the color may be, what treatment will our artistic friends suggest that can so well harmonize the woodwork with the oak furniture (and possibly the oak floor) as to grain the room to correspond with the furniture? A recent writer says to use a scumble or oak-graining color, but to attempt no figures on the work. Why not have some suggestion of the grain of the wood as well as the color? The effect of the color is to deceive, and is it any worse to add the figures?

A book may be composed of sheets of blank paper, and,

as Lord Byron wrote, "A book's a book although there's nothing in't." There is an element of romance in graining. The oak color alone on a room would be as a book of blank paper to the novelist, while the grained room would tell a story and either be true to nature or border on the romantic, according to the skill and ability of the grainer.

Again, suppose for certain reasons it is found necessary to place iron doors and frames in the corridors or halls of a building; if the surrounding woodwork is finished in natural wood, how will we manage to make the iron doors correspond with the other doors and the adjacent woodwork in any other manner so well as by properly graining them? Must we paint them black or a "wood color" because, being made of iron, it would be "false in art" to make them appear as wood?

There are things often seen in architecture that are as false as graining. This would not justify the use of graining; but some of its critics forget their own shortcomings in their haste to follow the rules laid down by their teachers, and condemn graining because it is "false."

The near future may bring us doors made of wood-pulp, compressed, and with all panels, mouldings, etc. in one piece. According to our artistic friends it would be wrong to paint these doors in imitation of wood. Yet I believe that such doors will be made and many of them will be grained.

---

[1] Clipping from *Boston Herald*, 1914, says:

"Newspaper Row is getting an illustrated lesson on our modern method of erecting an office building and then hanging on it what seem to be its supports."

From "Essays and Memorials," by John W. Simpson:

"Architecture is no mere ornamentation which can be applied to an unsatisfactory structure in order to beautify it. It is either inherent to the composition or irremediably absent."

CHAPTER III

EMINENT GRAINERS OF THE LAST CENTURY

AMONG the earliest of the celebrated grainers of the last century who have passed to their reward, perhaps it would be well to mention Hay of Edinburgh as among the first. He was an artist in his day and generation, and brought the art of graining to a high plane. He is reputed to be the inventor of steel graining combs. Leather and horn had been previously used. He was an all-round artist, and decorated some of the finest buildings in Scotland. My respected friend, the late John Smith, a trade grainer of New York City and an excellent workman, having served his time in Edinburgh and later worked in some of the best shops in London, wrote me the following anecdote of Hay.

It seems that Abbotsford, the residence of Sir Walter Scott, was being decorated by Hay (probably between 1815 and 1825). I will give Mr. Smith's own words: " Hay of Edinburgh, where McPherson of Boston served his time, was perhaps the most scientific painter that Scotland ever produced. He was Sir Walter Scott's painter when the latter was in the height of his fame. You may have heard the story of him and Scott. Hay had painted the library at Abbotsford, ceilings and walls, etc., in imitation of oak, to match the bookcases, but before Scott came home he gave the walls a coat of kalsomine. Everything pleased Scott except the library walls, but perhaps he could suggest nothing better. Hay suggested making them to match the woodwork and bookcases — just the idea — so when the " Wizard " was gone, the kalsomine was washed off, and in the morning Scott was astonished, and when it was explained to him, he said it was as good as anything in his books."

Apprentices of Hay's were as a rule excellent workmen, and many of them were known as artists in their profession in whatever portion of the world they had settled. The late William J. McPherson of Boston, and his brother Walter, both of whom were grainers to the trade in Boston, Massachusetts, in 1854, were graduated from Hay's shop.

Bennett and Bogle of Glasgow were known as excellent grainers and marblers. The late William Munro Ross, a trade grainer of Boston, who died in 1878, and an artist of no mean ability, was one of their apprentices. His son, William M. Ross, also an expert trade grainer, died March 2, 1905. The firm of J. B. Bennett and Sons, successors of Bennett and Bogle, is still known for the excellence of its work in the specialties of graining and marbling.

Undoubtedly the late Thomas Kershaw of London, England (born at Standish, near Wigan, Lancashire, in 1819, died at London, 1898), has done more for graining than any one man of modern times by bringing it to the attention of intelligent people as a pleasing form of decorative art; and by his exhibitions of graining and marbling at the first great exhibitions of the world's industries in London, 1851, Paris, 1855, and London, 1862, he exemplified to the world that such work was worthy of a place in the most artistic residences. Mr. Kershaw, when but twelve years of age, was apprenticed for nine years to Mr. John Platt of Hall Street, Bolton, Lancashire, his father paying £23 (or $115), a large sum in those days, for the privilege of having his son taught the mysteries of painting and decorating. No eight-hour law was in force in those days, and it is said that after working for ten or more hours a day for his master at such drudgery as the apprentice of his time had to do, he found time to work two, three, or four hours extra in his own room, studying the grains of woods and endeavoring to successfully imitate them, and so well did he succeed that long before his apprenticeship expired his fame as a skilful grainer had reached beyond his own shire.

His example should be an incentive to all young workmen who aspire to be grainers. Study the *wood*. Imitate no man's work. In a letter to me, written by Mr. Kershaw in 1893, he says : —

"I never studied another man's work or his method of working. I went direct to nature. She has been my only schoolmaster, and anything in the shape of woods and marbles this side of Heaven or the other place I determined to tackle and make a business of it, but for the last thirty-eight years left it off and commenced house painting, having served nine years to the trade I felt more entitled to this than a lot of counter-jumpers who are not practical except in lying. I find your prices for graining are reasonable and compare favorably with prices here for ordinary work, and if you work at high pressure, steam-engine rate, you should earn good money. Graziers nowadays work with their coat on and take life much easier than in my graining days. I will not say what I used to do, you would not believe it. Should you ever come, I will tell you that I found no royal road to fame. If you are an enthusiast, full of energy, take a trip to this small city, call upon me, and I will show you a large room full of specimens of my own work done before you were born. I received the Painter's Magazine you sent me and am much obliged.

"From yours truly,
"THOMAS KERSHAW."

Mr. Kershaw was awarded medals at the London Exhibition of 1851, Paris, 1855, London, 1862, and in 1860 the Painters-Stainers Company gave him the freedom of the city of London, an honor seldom conferred upon a practical grainer and marbler. Let us hope that the youth of to-day will by persistent efforts strive to attain the heights he succeeded in scaling.

The late Mr. John Taylor, of Birmingham, England, was

a keen rival of his elder contemporary, Thomas Kershaw, and I have heard it stated by men competent to judge that Mr. Taylor's marble panels excelled those of Mr. Kershaw. Both men worked from natural woods and marbles and cut out new paths in methods and tools, and each in his way was known and respected for his ability in the three kingdoms. Mr. Taylor was born in Glasgow, Scotland, in 1830, and died in Birmingham in 1901. He was early apprenticed to the painter's trade. In serving his time he worked at scene painting, and later, on reaching London, he worked for the trade as a grainer and marbler. He received medals for his work at the great exhibition, London, 1851, 1862, 1870, and in Paris, 1887, and in 1898 was given the freedom of the city of London. He settled in Birmingham in 1860 and did much for the artistic welfare of that town. He was a skilful picture painter and a member of the Royal Birmingham Society of Artists. His work had been hung and sold in the London Academy and in other large cities.

A friend of mine competent to judge assured me that the mantle of Kershaw had fallen on Taylor, and I doubt not that at the time of his death he was considered the greatest exponent of graining and marbling in the United Kingdom.

Would that the best grainers of this century would emulate their example and give ocular evidence wherever opportunity offers that skilful graining is not wholly a lost art. An object lesson in graining done by an expert is more convincing to the beholder as to the decorative possibilities than all the arguments that are arrayed against it.

Doubtless any of the larger cities of Great Britain or Ireland in the years from 1850 onward possessed skilful and accomplished workmen as grainers and marblers. I have heard personally of many such, but the names of Kershaw and Taylor deserve special recognition, because from youth to mature age they never failed to champion the cause of skilful graining, and at no small expense to themselves prepared and exhibited to the public such wonderful imitations of wood and marble as justly to gain for themselves the title of artists in their chosen calling.

Their work also acted as a stimulus to the efforts of those who were faithfully endeavoring to imitate nature in this special direction, however feebly; hence the merit of their work had a twofold value: beautiful to the beholder in or out of the trade, and especially stimulating to those within the trade, as it revealed the possibilities of the art; "for whatever man has done, man may do," and the humblest of us can be helped and encouraged by the work of men whose God-given skill we may never hope to fully attain.

CHAPTER IV

GROUND-COLORS

ONE of the first essentials for a successful imitation of wood is that the ground-color, or foundation on which the graining is applied, should be similar in color to the wood it is desired to represent.

Too little care is given to this feature of the work. The grainer is often asked to do dark oak on a light oak ground-color. Many an excellent piece of work has been rendered ineffective by the use of a ground-color unsuited to the wood to be represented.

So strong is tradition that excellent workmen will often adhere to a color scheme which could be vastly improved, simply because in their early years they were shown by their masters that certain colors and proportions must be used to prepare ground-colors for graining certain woods.

These remarks are not made to condemn the work of the older school of painters or grainers, but it is a fact that few of these men saw good examples of the wood they were asked to represent and hence were liable to mistake in preparing the ground-color.

Undoubtedly the first requisite for good work is a properly mixed and properly applied ground-color. Without this it is impossible to produce a good job. The professional or trade grainer is frequently asked to "do a good job" on a ground-color that makes good work impossible. The primary principle, the shade or tone of the ground-color, has been neglected. It is extremely difficult for the most expert grainer to even approach the color of the wood in his work unless some effort has been made to have the ground-color similar in tone to the lightest part of the wood to be represented; yet many painters pay little attention to this feature of the work and then blame the grainer for an imperfect job.

One general principle for preparing ground-colors may safely be laid down for any sort of wood: the color of the ground should be similar in tone and, if anything, slightly lighter than the lightest color to be seen in the wood it is desired to represent — not the unfinished wood, but wood filled, shellacked or varnished, and finished.

In matching the doors of a room to the standing finish of natural wood it would be better to first complete the standing finish, or at least up to the last coat of shellac or varnish, before graining the doors. Then remember that, as a rule, the grained work on interiors of dwellings often retains its original color for years, while the natural wood invariably turns darker by absorbing the oil from the varnish and by exposure to light and air.

It is not uncommon to see grained doors which, when first done, matched fairly well in color with the natural-wood finish by which they were surrounded, but when done for a dozen or more years, they frequently appear several shades lighter than the natural wood, for the reasons above stated. Bearing this in mind, it is wise to

GRAINING, ANCIENT AND MODERN

have the color of the grained doors, when finished, slightly darker rather than lighter than the wood to be matched. On outside exposure nearly all hard wood turns dark, especially if it receives frequent coats of oil or varnish.

If the work to be done is on new wood and the wood is clear white pine, spruce, or whitewood, the work may be prepared for light graining by giving it two thin coats of white shellac; or a coat of white glue-size may be given it, although it is not recommended, as it raises the grain of the wood and can only be used successfully where the graining is to be done in oil colors, as the glue-size would soften up and mix with the water-graining color, thus allowing no proper foundation for the work.

In preparing the ground-colors for graining use nothing but the best of finely ground white lead and colors ground in linseed oil. Any time or trouble expended in the careful preparation of the groundwork is amply repaid in the improved appearance of the finished work, and good work cannot be done without pains being taken at each stage of its progress. After mixing the color it should always be strained through fine muslin. The manner of thinning the ground-color will depend on the condition of the work. If the wood is new, only a small amount of turpentine will be required in the color for the first coat. Such work ought to receive three thin coats, but as a rule it is prepared with two coats on ordinary work. Always use plenty of dryer in the priming coat. The second and third coats should be thinned with not more than one part linseed oil to three parts spirits turpentine. If a little good varnish is added, it helps to harden the work and leave it in better condition for graining.

If the work is to be done over old paint, a thorough sandpapering is necessary. If the old work is badly cracked or chipped, all the old paint should first be removed; it costs a little more in the beginning to do this, but it is worth while in the end.

There are many patent removers in the market, but the old-fashioned compound of lime and soda will often remove anything of an elastic nature, but for old hard paint it is not sufficiently caustic. Beware of removers having either bisulphide of carbon or wood alcohol in their formula, as both are dangerous to health.

The burning lamp or torch if carefully used is often quite effective in removing old paint, and it has the merit of leaving the work ready for the brush after sandpapering.

In case the paint is removed by the torch the thinners for the priming coat should be about one-half raw linseed oil and one-half spirits of turpentine with plenty of dryer, as the torch will have drawn out most of the sap from the wood, leaving it very dry. The excess of turpentine in the priming coat is more likely to be drawn into the wood. Too much oil might cause trouble later.

If old work is to be repainted and grained without removing the old paint, it should first receive a thorough washing with a weak solution of washing soda — about one pound of crystals of soda to three gallons of water. Rinse off this solution with clean water to which has been added one-half pint of vinegar to the gallon. This will neutralize the action of the soda if any should be left on the work.

Sometimes a job is done on one coat of groundwork over old paint or old graining, but it is unwise to do this. If it must be done, first touch up all bare places, dents, bruises, etc., with quick color mixed rather thin, and in applying the ground-color don't try to put on too much paint at once.

The grainer would rather have a smooth surface fairly well covered than a rough, ropy one, however well covered. Then again ropy or rough groundworks always show up very badly after the varnish has been applied. Don't forget that smooth work cannot be done unless care is taken with each coat and sandpaper has been used between coats.

It is customary for the grainer to lightly sandpaper the ground-work before applying his color; but on very light work this must be very carefully done or omitted altogether, for if any scratches are made in the ground-color, they will show in the finished work on such light colors as maple, satin wood, etc.

The use of red lead is recommended for mixture with dark ground-colors where the old paint is not removed. It helps to bind the color more firmly to the old paint. The color should be kept stirred to prevent settling.

In some cases a badly cracked foundation can be helped by using color thinned, about half linseed oil and half spirits of turpentine with plenty of dryer added. Thicken this color with bolted whiting and after coating in a panel rub it with a block of pumice-stone which has been faced smooth on one side with an old rasp. Draw the pumice-stone down the panels after having thoroughly rubbed to the work. This will leave the crevices and cracks filled with the thick paint and fine particles of pumice-stone. It also helps to amalgamate the new paint with the old. Do not use the brush to smooth up, as it will draw the paint out of the cracks. Use a scraping knife and leave all the color you can in the cracks. When the work has dried thoroughly, sandpaper and apply two coats of ground-color mixed in the ordinary way. For a hurried job this method will be found useful.

Many of the washable distemper paints are recommended by their makers as being suitable for groundworks for graining; but their use is not recommended, as they often contain soda or other alkaline matter that is destructive to all subsequent coatings of oil color, both graining and varnish. Then again, if moisture attacks the work from beneath, it is likely to cause the water-color paint to loose its hold and force it and all succeeding coats away from the wood.

CHAPTER V

GRAINING COLORS

AFTER many years of careful investigation the writer has for several years used only the finest dry colors in mixing his graining color. (Burnt sienna, Vandyke brown, and black are excepted.) There are several reasons for this. First, he can use his own formula for mixing the proportions of each color. The adulteration of dry colors can be more readily ascertained than if the colors were ground in oil, and a recent advertisement declares that colors can be ground dry as fine or finer than in oil. Then the color is never fatty nor are there ever any skins of color in the pot. If there are any coarse particles in the color, they are deposited in the bottom of the pot. Another advantage is that the work can be done in either oil or water color from the same base. The mixture of the various dry colors being carried in one receptacle, where the work is largely of one kind of wood the prepared color has only to be thinned to be made instantly ready for use. If cheapness were to be considered, this also would recommend this method. But it is not for this reason that the writer has adopted this plan, but chiefly because he can readily ascertain the purity of his colors, knowing that they are not ground in fish oil nor petroleum, and he can produce certain mechanical effects by their use that are not so readily obtained where colors ground in oil are used.

The reasons for using burnt sienna, Vandyke brown, and black, ground in oil or in water, are that the first two are naturally gritty and require grinding and the latter is in more compact form as well as being smoother from being ground in oil. A trifle of dryer is usually added in grinding it, so that it dries more readily than if ground wholly in oil.

CHAPTER VI

THINNERS FOR OIL COLORS

IN referring to thinners hereafter for oil colors, we make a compound as follows: Two-fifths of a gallon of raw linseed oil, three-fifths of a gallon of spirits turpentine. Add to this one-half pint good liquid dryer and about two-thirds of an ounce of yellow beeswax. Cut the wax into shavings and melt in a tin can over a gas jet or on the fire. When the wax is melted, take from the fire and add slowly about one pint of turpentine and pour while warm into the thinners previously mixed.

The beeswax may be dissolved without heat by cutting it into shavings and placing it in a wide-mouthed bottle. Fill two-thirds of the bottle with turpentine, and allow it to stand over night in a warm room. A few violent shakes of the bottle in the morning will complete the dissolution of the wax. It can then be added to the thinners and the whole well shaken up. Care must be taken in cold weather to have the wax completely dissolved in the color, so that it will not float on the top of the pot, otherwise it will cause serious trouble and retard the drying of the color for an indefinite period.

On new work, as a rule, slightly more oil is necessary than on old work, and for walnut or cherry in oil less oil is required than for oak or ash.

Thinners for water colors may be clear water (if the colors are ground in any binding medium) or one-third to one-half stale beer to two-thirds or one-half clear water. If beer cannot be obtained, one-third vinegar to two-thirds water, adding a little sugar, will make a good thinner, or you can use skimmed milk. As a rule too much beer is used in the color, and strong beer is likely to cause cracking of the varnish.

CHAPTER VII

MEGILP FOR OIL COLOR

IN imitating many woods, especially in oil color, we need something more than the colors and thinners to prevent the work from flowing together after being combed or wiped out; also in water color to prevent the color from drying too quickly. Such preparations are known as megilp. They alter the density of the color without altering its shade and allow the combed or wiped-out work to remain just as you leave it. As a rule too much megilp is used and it causes the work to look stringy or ropy. For many kinds of work, where too much oil is not used, we can dispense with megilp altogether, especially if we use finely prepared dry colors. An extra quantity of dryers acts as a megilp. In my opinion it is unwise to use more than one ounce of megilp to the gallon of oil color, and it should always be added to the color while warm, after being dissolved in turpentine.

In a book entitled "The Painter's and Glazier's Guide" by Whittock, published in London, England, in 1832, a recipe is given for megilp for graining color which seems to have been a standard, as I have seen similar recipes in later publications. The formula is this: "Take eight ounces of sugar of lead and eight ounces of rotten-stone, grind them together as stiffly as possible in linseed oil; then take sixteen ounces of white beeswax, melt it gradually in an earthen pipkin, and when it is fluid, pour in eight ounces of spirits of turpentine; mix this well with the wax, and then pour the contents of the pipkin on the grinding stone to get cold. When cold, grind the rotten-stone and sugar of lead with the wax and turpentine, and it will form an excellent megilp, which if kept in a jar with a mouth wide enough to admit a pallet knife, and well secured from dust, will keep almost any length of time and always be fit for use. When any of the megilp is required, take out a little at a time with the pallet knife, and if it is too stiff, soften it with a little boiled oil."

It is doubtful if many of our modern workmen go to the pains of preparing their megilp in such a thorough manner, yet it is worth while to do so. The grinding into the megilp of the sugar of lead as a dryer is an excellent plan, the beeswax being a very slow dryer, and often when added to graining color in cold weather it curdles and floats on the surface, and if not thoroughly incorporated with the graining color will seriously retard the hardening and drying of the work.

MEGILP FOR WATER COLOR

When the color dries too quickly or is not sufficiently thick for special work, it may be megilped or thickened and its drying retarded by a solution of soap, wax, and turpentine ground in beer and added to the color. Use very sparingly in the graining color, as it clogs up the tools and has little virtue except to allow opportunity for combing, on account of drying much more slowly than the beer and water vehicle ordinarily used; one-half ounce of soap cut into shavings and dissolved in one-half pint of hot water, one-half ounce of wax cut into shavings and dissolved in one-half pint of hot water and one pint of stale beer, may be mixed together on a grinding slab a little at a time and kept in a wide-mouthed bottle tightly corked. The mixing may be accomplished by heating the beer, and after pouring the hot water and wax into the hot soap and water and thoroughly shaking, add the hot beer to the mixture and continue shaking until all the ingredients are thoroughly incorporated; when cool it ought to assume a jellylike consistency. Care must be taken that it does not rub up in varnishing; try several experiments with it before you adopt it for regular use.

CHAPTER VIII

TOOLS

IT is astonishing how few tools are used by some of the most experienced grainers to produce the results which are apparent in their work. A few old combs, a piece of rag, and the necessary brushes, colors, and thinners are all that they require to do most excellent work. The first requisite is undoubtedly the natural ability to transmit to the painted work the true figure of the wood they desire to represent. There can be no doubt, however, that good tools, and many of them, allow the beginner better scope and assist him to produce better effects than without their aid. The adjacent pages will show illustrations of all the tools necessary to imitate any ordinary wood.

The rubber comb is made from a piece of rubber with cloth backing, which can be purchased at any shoe-findings house. It is made to apply to the soles of rubber boots. It is sold by the pound, and one pound will make four or five combs.

The selvedge of a piece of straw matting makes an excellent fine comb if cleanly cut on the ends. It will do moulded work most effectively on account of its flexibility.

The steel combs are generally made in Sheffield, England, and are often given special treatment by the grainer, as shown in the illustration.

The check roller was invented by William Jones of Manchester, England, and is very effective in transferring an imitation of the pores of the wood which have been filled dark. On the lighter shades of oak it is seldom necessary to use a check roller; the color is transferred from a brush held over the zinc disks, the roller being pushed forward.

The long-handled pencil brush is used for putting in

GRAINING, ANCIENT AND MODERN

heart grains as well as quartered grains. The other fitch tools and sash tool are used for smaller work.

WATER-COLOR KIT

The sponge and rubbing-in brush play important parts in this process. The long-haired stippler is used chiefly for walnut and mahogany. The mottlers, blender, etc. are for use in maple, mahogany, walnut, etc. The overgrainers are the last tools used in either oil or water colors.

Combs made of cork are sometimes used for oil graining, and they do excellent work; the teeth may be notched on the edge of the comb, or they may be cut through as shown in the rubber comb. If notched teeth are used, the comb must carefully be held at the proper angle, so that the color is divided by the notched teeth.

CHAPTER IX

RUBBING IN OIL COLOR

JUST why the application of the graining color is called "rubbing in" it is difficult to determine. Possibly in the early days of oil colors the graining color was mixed so thick as to require a great deal of "elbow-grease" to spread it on the work, and hence the name rubbing in.

After the oil color has been properly mixed, a proper brush is carefully worked into the color by being dipped deep into the pot and scraped on the edge of a hard-wood stick which is kept in the pot to stir the color.

A flat brush or a good oval one is preferable to a round brush. The bristles should be three to four inches in length, and with such a brush several panels may be done at one dip. A sash tool is often necessary, but it is a loss of time to do work with a sash tool that can be done with greater ease and more despatch with the larger brush. The professional "rubber in" rarely has a sash tool in his hand. Life is too short to do edges of doors and frames with a sash tool when the large brush will do them equally well. An edge-to-paper can be cut as successfully with a proper flat brush as with any sash tool, and with greater command of the brush. Your hand is less liable to slip with the large brush than with the smaller one.

The average painter who rubs in for a grainer seems afraid to dip his brush into the color more than one-half an inch, while the old hand takes pains to take a "full dip," so that his brush will not dry up in the heel and will be more flexible than if the heel of the brush is kept dry. Where a whole room is to be rubbed in, it is needless to take half a dozen dips to rub in a door side if it can as easily be done in two or three.

In rubbing in the graining color on an ordinary door side we first take a good dip of the color in the flat brush and apply it to the mouldings of the upper panels of the doors. If there is still plenty of color in the brush, do the edges of the door and the mouldings of the lower panels, then go back to the upper panels and finish them, taking up any surplus color from the mouldings which were done with the full brush. Never take a second dip while there is color in the brush. Two, or at most three, dips ought to suffice for an ordinary 2 ft. 8 × 6 ft. 8 door.

In rubbing in the tops of doors or frames do not carry up the pot on the step-ladder. It is better to avoid the risk of upsetting it, and a yard square of work can be done with one dip if necessary. Some grainers prefer to have two rubbing-in brushes, an old one and one just well broken in. By taking a dip in each a lot of time is saved and the color can be applied on the tops of doors, casings, etc., without so frequently returning to the pot.

Graining color is spread more with the sides of the brush

than with the ends. Many men use the rubbing-in brush as a scrubbing brush should be used, wearing the brush away on the ends of the bristles. If the color is properly mixed, its application is not a difficult task unless there is an improperly prepared ground-color to work on. Allow the arm free play. In rubbing in a piece of sheathing running horizontally, a yard or more in length of each board ought to be done and care taken to show no laps on the work. One dip ought to rub in several boards.

The full brush should be applied to mouldings so as to thoroughly cover all crevices. The surplus color can easily be removed with the dry brush later on. No man who is a careful workman will, after taking a dip, draw his brush over the edge of his pot to remove the surplus color. One or two taps of the brush against the inside of the pot will remove the surplus color more quickly and effectively without leaving streams of color on the outside and inside of the pot or spatters of color on the floor. The same rule holds good for all kinds of painting, but particularly to color or varnish that is very thin. Neatness of person and cleanliness of work often go together.

Unless the color is properly applied the grainer is handicapped in no small degree. It is a mistake to suppose that any one can apply the graining color. It takes years of practice to do this successfully and rapidly. Old grainers have a saying that none but a grainer can properly rub in for a grainer. It is amusing to watch the struggles of the amateur in his attempt to apply the color evenly. He often gets enough color on his brush to rub in several panels, but he tries to make this amount of color appear clean and smooth on one panel, and the more he works over it, the worse the panel looks.

Graining color should be rubbed out evenly, not too dry but always clean and free from cloudy places.

If the groundwork is glossy from the use of too much linseed oil, the graining color is, in cold weather, likely to crawl or creep and not adhere to the work except by the use of considerable elbow-grease.

An effective remedy for this trouble is found in benzine, which can rapidly be applied to the work, using a clean pot and brush, and wherever it touches the groundwork the crawling is completely prevented. This preparatory work can, if necessary, be done hours before the work is rubbed in. A little bolted whiting should be added if the work is to be done in water colors.

RUBBING IN WATER COLOR

If the groundwork has been properly prepared, there should be no difficulty in applying the graining color with a large sponge, or with the flat brush.

If too much oil has been used in the paint, the work will appear glossy, and it will require much more effort to make the color adhere to it.

If the crawling or cissing is too pronounced, it may be remedied by the process previously described for oil colors, or the sponge may be dipped in a tin of damped whiting, or a cake of soap may be held in the left hand and the sponge or brush rubbed in it, which will help to stop the crawling when the sponge or brush is rubbed to the groundwork.

In rubbing in doors in water color it is wise to leave the long stiles or rails until the rest of the door is finished. Take a sheet of coarse sandpaper; give it a good coat of shellac on the smooth side. This will prevent the absorption of color. It can then be placed against the mitres of the door, rough side in, which will prevent it from slipping, and with a sponge wet in clear water cut off the color clean to the cross stiles; then rub in the long stiles, or rails, and finish the door.

CHAPTER X

COMBING IN OIL COLOR

AFTER the graining color is rubbed in, the portions of the work that are intended to be left with plain grains or that are to be figured in quartered oak are combed; that is, a background representing the pores of the wood is produced by using rubber or steel combs. If the grain to be represented is yellow pine, ash, chestnut, cherry, cypress, or walnut, a steel comb covered with a rag will do good work, or a rubber comb cut with teeth about five to ten to the inch can be used without being covered with a rag. Some grainers prefer to take a coarse steel comb which has been placed in a vise and the corners of the teeth rounded with a file so that a comparatively small surface of the tooth of the comb comes in contact with the work. This leaves wider spaces between the teeth; but by pressure on the comb, the teeth of which are covered with a rag, the width of the lines left by the teeth is made greater. As the comb wears away the round corners of the teeth are lost and it is necessary to again file them.

Remember that the color is to be pushed aside by the rubber comb, and the track left by the comb need not indicate that all the color has been removed from the work. The color should be gathered in the dark lines, but it must not be too thick or it will stand out in ridges, which is just the opposite from the natural effect it is desired to imitate, as the dark lines in wood are most frequently open pores which, when filled, outline the grain with greater distinctness, and, as a matter of fact, they are below the level of the surface unless thoroughly filled. They are never above the surface; hence the grainer must be careful not to have his work appear ridgy, which would be the case if the color was too thick and left ridges when combed.

If clean rubbing in is essential to the looks of a job, clean combing is doubly so. Many a job would be vastly improved if the grainer had taken pains to keep his combing clean and distinct. A cleanly executed job of combing is as good an imitation of wood as is the most elaborate piece of figured work.

The beginner cannot spend too much time practising combing, always taking a piece of natural wood for his copy, and unless he can master this detail it is hopeless to look for success as an all-round grainer.

The professional, or trade grainer, seldom carries an extensive set of combs. Often two or three old steel combs and a few rubber ones, and possibly the selvedge of a piece of Canton straw matting, comprise his kit for oil graining. With these few tools, and some cotton rags, an expert can do excellent work. It is not intended to condemn a larger kit of tools. It is sometimes necessary in order to execute special work to have special tools, but the tools named above are all that are absolutely necessary for combs.

After wiping out heart grains with a rag it is advisable to carry out the fine lines on the outside edges of the heart grains by using a rubber comb about one and one-half inches wide, the teeth of which are cut rather deep and about ten to the inch. Cover the comb with a thin piece of cotton cloth and draw through the color, closely following the last line wiped out by the rag. Considerable pressure must be used or the lines will not appear as distinct as those made by the thumb covered with the rag. It is possible to use the comb for this purpose without first covering it with the rag, but it is safer to keep the teeth covered. A coarse steel comb may be used in this manner if the teeth have been filed as previously directed.

It takes years to acquire an expert touch in combing,

GRAINING, ANCIENT AND MODERN

and it is useless to attempt the finer lines of work until a fair success has been achieved in combing.

Great care must be taken to wipe the teeth of the comb clean every time it is used or the work will suffer in consequence. Hold a rag in the left hand, and keep the teeth of the comb clean, wiping the teeth on the rag every time the comb is used. A selvedge of coarse Canton straw matting makes an excellent comb for the finer grains of oak or any close-grained wood; cut a piece about four inches wide, leaving the selvedge for back of comb. Let the matting be about two and one-half to three inches in length, and have the edge cut clean with a sharp knife or scissors. This makes a very flexible comb, which readily adapts itself to moulded work, and more work can be done with it on mouldings at one sweep than by any other tool except a corn broom or a water-color overgrainer. It is quickly worn out, but its cost is very small. New matting makes the best comb, but old matting is not to be despised. Some of the coarse matting of woven grass that is wrapped about tea chests or other articles that are imported from the Orient makes a good comb for certain purposes. A handful of rattan shavings can also be used or a bunch of cotton twine. Many grainers cover their fingers with a rag and make coarse combing by this method, but wood is seldom successfully imitated in this way.

Some grainers prefer to cut notches in the edge of the cork or rubber comb rather than to cut the teeth through the edge; in this case the comb must be held at the proper angle, in order to properly comb the work.

In combing a background for quartered oak a careful study of the wood will show us that the grain appears to consist of a series of pores of longer or shorter length, and the spaces between the pores is of different widths, often varying in the same board from very wide to very narrow. A graduated rubber comb with teeth of different sizes is useful for such work.

We first draw the rubber comb through the graining color, holding the comb at a slight angle so that the graining color is gathered and left in a track from one edge of the teeth of the comb. A wide-toothed rubber or cork comb may be used in beginning to comb a wide panel. After two or more sweeps of the comb a finer-toothed one may be taken, and the lines made by the coarser comb closely followed. If the panel is very wide, a still finer-toothed comb of the same material may be used until the other side of the panel is reached. Be careful to follow closely the general direction of the lines made by the first comb.

When the panel is filled with parallel lines made by the rubber or cork comb, they should be broken up into smaller lines and made to represent the pores of the wood by judiciously using the steel combs, drawing them through the lines made by the rubber or cork comb and at a slight angle so that the effect of the pores is obtained.

By taking a medium-toothed steel comb and breaking out every third tooth a more woody effect is produced in one combing than if the teeth were not removed. The finest steel comb may be treated in a similar manner and used to make the interlocked effect of the finer plain grains by using it over the tracks made by the fine rubber or cork comb. The work is then ready to wipe out the quartered grains.

When oak heart grains are wiped out with a rag, they can be made a great deal more like the grains of wood by using a blending comb. This is an ordinary two-inch medium steel comb and is used as follows: break out every other tooth; then heat the teeth red hot, place the comb in a vise, and bend the tips of the teeth backward until they are at right angles with their original position, but with a rounded corner. When cool, cover the bent teeth with a thin piece of rubber such as toy balloons are made of and draw it tightly over the teeth. This will leave hollows

GRAINING, ANCIENT AND MODERN

between the teeth, and the comb can then be drawn against the edges of the heart grains made with the rag, and the edges are serrated and slightly blended, giving a much more natural effect to the work. A similar comb, without bent teeth, and without being covered with a rag, may be used for the same purpose, but the work lacks the blended effect given by the rubber-covered comb. If the thin rubber is tied around the end of each tooth of the comb, it will do good work as a blending comb without having the teeth bent.

COMBING WATER COLOR

For combing in water color, rubber and steel combs may be used, also combs cut from a potato or a turnip. It is necessary to add a little beeswax and soap or glycerine to water color if successful combing is attempted. The beeswax may be melted by boiling it in water. Be careful not to get in too much; a little goes a great way. The water color should not be allowed to get too cold; a little alcohol added to the water helps to assimilate the wax and keep it in solution. The megilp compound previously described should be used if necessary.

The overgrainers can readily be made to take the places of rubber combs for water-color work, and the steel combs can be used over them while the work is wet. A stippled effect produced by the use of the long-haired, grainer's stippler is often an excellent background from which to work up quartered oak in water color, provided the stippling is not done coarsely nor of too dark a color.

Jones's patent graining check rollers may be used with excellent effect in either oil or water color. For some sorts of dark oak in oil the figure may be wiped out without combing, and the effect of the pores of the wood may be obtained by use of the check roller used as an overgrainer after the work is dry.

The check roller is made of thin disks of zinc, which are

notched on the edge like the teeth of a comb. The centre of the disks have a hole punched through them, and they fit loosely on a wooden shaft through which runs a steel pin. Between each disk a brass ring is placed on the wooden shaft. Its thickness determines the width of the spaces between the disks. A handle of thin steel is sprung over the ends of the steel pin which protrudes through either end of the wooden shaft, and its own tension holds it in place. The disks revolve as the roller is pushed over the work, and the color is fed to them by a brush held firmly against the disks as they revolve. Care must be taken not to use strong vinegar on the disks nor to clean them by soaking in potash, as either treatment will dissolve the zinc and sharpen the disks to a knife edge.

Wooden rollers have been made to do similar work, but they fail to adjust themselves to the inequalities of the work and only touch it when perfectly level; hence they do not equal Jones's check roller.

The check roller can be used in water color on the groundwork before the graining color is rubbed in, and the marks which appear on the light portions of the work when the color is removed to make the figure can be rubbed off with a damp cloth.

For the lighter kinds of oak it is unnecessary to use the check roller except as it may be used to put in the medullary rays which appear in the heart grains. These may be put in by using either oil or water color on the work previously grained. On the whole, the best effect is obtained by the use of water color.

CHAPTER XI

GRAINING CRAYONS

FOR imitating certain kinds of wood, crayons or crayon pencils are of great assistance. The latter can be bought at almost any art-goods shop, and the former can be made to suit the requirements of individual cases.

In making crayons for water-color work the base of the crayon is composed of pipe clay or china clay. Bolted whiting will do, but it is likely to be more gritty than the clay; it also has a bad effect on siennas and umbers, making them fade. Make the crayons on a marble slab or on glass. When made, lay aside to harden. The colors for tinting the crayons depend on what wood you wish to represent. All colors should first be finely ground in water. For making crayons for light oak or ash, proceed as follows: Dissolve about a teaspoonful of gum arabic in two tablespoonfuls of boiling water. Mix this with about two tablespoonfuls of pipe clay. Then add the colors, which should contain as little water as possible. Use stale beer to thin the paste color if it is too stiff to work freely. One part burnt umber to two parts raw sienna will make a good oak or ash crayon. If a dark oak is desired, add Vandyke brown, or a little drop-black, till the desired shade is obtained. If colors ground in water cannot be had, use any colors. A small percentage of glycerine added to the mixture will make the crayon work smoother. It is necessary to have the groundwork flat or nearly so or the crayon will not mark on the work.

For cherry or mahogany use burnt sienna and a little burnt umber. For walnut, burnt umber and Vandyke brown; for rosewood, drop-black.

For overgraining bird's-eye maple the crayons in wood will be found much better adapted to the work than any made by hand. They should be of a light reddish brown, about the color of burnt sienna. Crayons can be used in oil-graining color, but as a rule are too soft for that purpose. They can be made tougher by the admixture of a small quantity of beeswax melted in turpentine.

As a rule the work done by crayons is apt to look harsh and crude unless carefully blended. Heart veins and the fine dark veins of quartered oak or sycamore are the only things for which the crayons can successfully be used. Some grainers soak the crayons in oil before using them in oil colors; for maple the crayons encased in wood are much the best.

Experiments with varying proportions of the above ingredients will determine the crayon best suited for individual needs or for special work.

CHAPTER XII

WIPING OUT HEART GRAINS IN OIL COLOR

THE heart grains of oak, ash, chestnut, or any open-grained wood may be represented by folding a soft piece of cotton cloth two or three times and placing the thumb nail of the right hand in the rag, holding the loose portion of the rag in the left hand and then drawing the outlines of the grains. When the color has become slightly set, or partially dry, the inside edges of the figures may be softened with the rag on the thumb, leaving the outline clean and sharp. A steel comb may then be used and drawn lightly against the outlined or sharp edges of the work and the whole slightly blended with the rubbing-in brush or badger blender, drawing the brush always against the sharp outlined edges of the figure; or the blending comb previously described may be used. Its advantage is that a round-edged heart may be made into a serrated one by the use of this comb.

If a pencil brush or fitch is used to make the heart grains for oak, the graining color should be rubbed in quite sparingly, and when slightly set the coarse steel comb should be used and drawn through the color, the comb being held at an angle which will allow the color to be deposited from one edge of the comb. The split steel comb (that is, a medium steel comb having every third tooth broken out) is then drawn across the edges of the lines made by the coarse steel comb, and when a sufficient portion of the color is removed a pencil bristle fitch is dipped in some of the graining color and the heart grains put in. The work is then blended with the rubbing-in brush or a badger blender and the edges of the grains are sharpened and the inside portions softened. Care should be taken to vary the direction of the figure of the work; endeavor to be versatile and work out the figure of the heart grains up or down in the centre or on the edge as the work may require.

CHAPTER XIII

CURLY MAPLE

Ground-color. — White lead, a very little medium chrome yellow with a touch of vermilion red. Thinners: one-third oil to two-thirds turpentine, adding a sufficient quantity of dryers.

Graining Color. — Crimson lake and drop-black with raw sienna, or raw sienna and raw or burnt umber; overgrain with thin burnt sienna.

Tools. — Sponge, rubbing-in brush, mottler, blender,

chamois leather, overgrainers, bone comb, and camel's-hair pencil or crayon set in wood.

Curly maple is a very handsome wood and is found growing in all temperate climates; from fifty to sixty varieties of maple are used for timber. It is often cut from the same tree as bird's-eye maple, the latter being the outer layers of the wood. It is better to represent this wood in water colors.

First dampen the work with a sponge wrung out of a mixture of one-third stale beer to two-thirds clean water. If the color "crawls" or "cisses," dip the sponge in some dry bolted whiting and rub over the work. This will effectually stop the crawling, and also act as a megilp. Rub in the graining color, which is composed of raw sienna and a touch of crimson lake and drop-black, or raw and burnt sienna with a touch of umber may be used; or the raw sienna may be entirely omitted and crimson lake and drop-black will be all that is necessary to use. It depends on how the wood to be matched has been finished. New wood will be light and show but little of the raw sienna shade, while old wood invariably takes on a yellowish brown shade, which is fairly well represented by raw sienna.

Have two separate pots: one full of clean water, in which the tools must occasionally be cleaned; and the other containing a small amount of graining color, just enough to do the work in hand. A little maple color goes over a great surface. Have a palette, a piece of glass, or an old plate, on which some of the thick color is placed and on which the overgrainer is left when not in use. The sash tool can be dipped in this thick color and a little taken up at a time and spread out with the large brush which is used in the thin color. In graining a door maple, the mouldings are always left till the last thing, and then neatly cut in and mottled if necessary.

When a piece of the work is rubbed in with the color,

GRAINING, ANCIENT AND MODERN 35

take the mottler, and after having wet it in clean water and shaken it thoroughly, proceed to mottle the work, taking off the color in irregular patches similar to the grain of the wood and blend lightly crosswise with the badger blender. Or the fingers may be dipped in clean water, and by drawing them across the work a mottled effect is produced which, when blended, is very similar to the figure of the wood.

After this work has become dry it must be overgrained very lightly, using the short-haired overgrainer and a very thin wash of burnt sienna, with which the overgrainer is charged, and the bristles separated by the bone comb, and the fine overgrain lines applied over the mottled work, giving the lines an undulated appearance, using great care to have each set of lines made by the brush follow exactly the same direction as those previously made. This work must be quickly blended crosswise, so that a sharp edge is given to one side of the overgrained lines.

Cut down the mitres of the long rails with a clean sponge and rub in slightly lighter (or darker) than the cross stiles. When all is dry, pass the hand lightly over the work and it is ready to be varnished. If a very deep shade is required, it may be given a very thin glaze of water color after one coat of varnish has been applied and the whole revarnished.

CHAPTER XIV

BIRD'S-EYE MAPLE

THE same ground-color, graining color, and tools as for curly maple.

Nothing among our light woods can compare in beauty with bird's-eye maple, and of recent years it appears to have grown more in favor, especially in furniture. It is found in all temperate climates and is a very beautiful and durable wood.

There are a dozen or more varieties of bird's-eye maple, and the eyes or dots in the work vary according to the species of the wood or the soil in which the tree was grown. A piece in my possession, grown in the mountains of Italy, is completely filled with the "eyes." Not a square inch on the board contains less than two or three of the "eyes," and in many places seven or eight "eyes" appear to the square inch.

It would not be wise to represent the eyes so profusely on the average job, for it would give it the appearance of being overdone. The general character of the wood places the eyes in clusters with scattered eyes between. Sometimes one side of a panel will be comparatively free from eyes and the other side will be filled with them.

Bird's-eye maple can best be represented in water colors, but it is not impossible in oil colors. It can be done more rapidly and with greater effect in water than in oil.

After having prepared the colors, as described for curly maple, rub in a panel, and after wringing out the chamois skin or a piece of soft cotton rag from a bath of clean water, form it into a roll and roll down the panel. This will take off the color in irregular patches. Blend at once with the badger blender, crosswise; or the mottler may be taken and portions of the color removed to show the high lights, and the blender used lightly. Or a roll of oil putty may be used and rolled down the panel, which will take off more or less of the color; blend at once. Or the backs of the fingers may be used and the color drawn into darker shades. Then use the mottler to draw up the color into little heads on which the eyes will later be placed. If the color dries before the work is completed, wet it over with clean water applied very carefully with a clean, short-haired overgrainer. A strong, high light generally divides the bird's-eye. Blend lightly and proceed with the work.

GRAINING, ANCIENT AND MODERN

When the first application of color is dry, the eyes should be put in, using burnt sienna very thin for the color. The amateur grainer generally puts them in by using the tips of his fingers in the wet color, but the wood is seldom represented in this manner. There are a variety of methods by which the eyes can be represented. A slice of raw potato may be cut so that it will take up the color like a rubber pad and apply it to the work in the form of a bird's-eye. Or a camel's-hair pencil may be cut, leaving about one quarter of an inch of hair from the quill. Then burn out the centre of the brush with a hot wire. The color can then be taken up and deposited on the work where the eyes are to appear. The eyes can also be made with the crayon pencil cased in wood similar to an ordinary lead pencil. This method can also be used for the overgraining, as there invariably appears to be a fine line of color of a darker shade which winds about the bird's-eyes. The eyes can also be represented by taking a soft piece of cotton rag, wet in the graining color, in which has been placed a little of the dark color and burnt sienna mixed together. Wrap this rag around a wooden skewer about the size of a lead pencil; fold the double edge of rag around the point of the skewer once only, winding the rest of the rag higher up the skewer. Keep the upper part of the rag thoroughly wet, which will drive the color down towards the point of the skewer, from which, by the folded edge of the rag, it can easily be transferred to the work and enough can be taken in the rag at once to do several panels. The size and shape of the eyes can easily be regulated by altering the shape of the rag at the point of the skewer.

When the eyes have been put in, the work may be overgrained — in some places very faintly, in others rather boldly, depending on the character of the eyes. The overgraining color should always be warmer in tone and considerably darker than the rubbing-in color, and when applied with an overgrainer or with a camel's-hair pencil, it should be blended to a sharp edge. Be sure to avoid the tendency to run over the eyes. This is never seen in nature. The eyes are always clean cut and distinct from the rest of the work and the overgrain invariably passes around, not over, them.

If the lights and shades are not sufficiently marked, any of the shadows may be darkened and the high lights thrown into greater relief by applying a thin coat of the graining color to the portions to be darkened and blending quickly. This must be done very carefully or it will look patchy and cloudy, and it should be done before the overgraining is applied.

Care must be taken not to overdo the work. Put the choicest grains in the panels and leave the stiles and rails with less figure. This rule will apply to nearly all woods.

An old, worn-out, flat brush makes a good mottler for maple. Saw off the handle close to the binding and wash thoroughly clean in strong soda water. It is then ready for use. The camel's-hair fine pencil may be used for overgraining, and oil color can be used for this purpose. If so used, it would be better to do it after the work had one thin coat of pale varnish.

Bird's-eye maple can be represented wherever it is desired to produce light and delicate effects, and for such purposes it is far superior to white paint or plain paint of any kind or color. The effect of a nicely executed job of maple is reposeful and quiet. It gives an impression of lightness and grace not to be obtained by any other means except by the use of the natural wood.

Maple is finished in many shades; but it is most beautiful when finished in white shellac or light copal varnish, leaving the mottlings clear and bright. It will assume a darker hue with age, and this hue is never satisfactorily imitated by staining the new wood. Where the work is to be matched to old maple, care must be taken to prepare the ground-color of the proper shade, which will be as nearly as possible similar to the lightest color to be seen in the wood. This rule for ground-color applies to all woods, light or dark. The grainer cannot produce the proper effect on a color foundation which bears no resemblance to the general color tone of the wood. It is better to have the color a little lighter rather than darker than the lightest shade of the wood, as the work can be overgrained or shaded to bring it to a darker shade; but unless the ground-color is sufficiently light the work will have to be repainted in order to produce a woody effect.

In our grandfathers' days drawing-rooms and chambers were often grained in maple with satinwood panels. The work of many old grainers, long since dead, is a monument to their intelligence and skill as craftsmen. To successfully imitate this wood requires a dainty touch and a deep knowledge of the figure of the natural wood combined with no small amount of technical skill. Nothing but persistent practice and study of the wood, using the best methods and tools, will suffice to acquire this skill.

The tools required to imitate maple need not be many. The human hand is a most wonderful tool and unaided can, if intelligently directed, do many things which would not be supposed possible. Mottling equal in effect to anything the mottler or cut tool can do may be produced by the fingers alone. This, of course, applies to water colors. Rub in the panel or stile, wet the hand in clean water, and to produce a mottled effect draw the fingers over the work from the palm of the hand in the direction you wish the mottles to appear. Blend immediately, and when dry overgrain. The knuckles of the hand may be used in a similar manner to produce the lights and shades in the panels previous to putting in the eyes. Nothing but practice will allow a person to become expert in this sort of work.

In making the eyes in the wood care should be taken to place the high light on the eyes in such a manner as to suggest the real wood; the eyes are seldom enclosed in a circle. The overgrain around the eyes is almost invariably intersected by the high light, so that the overgrain, as a rule, appears above and below the eyes and the high lights or mottle veins across the eyes. In some specimens the high light appears only on one side of the eyes and the stronger overgrain on the opposite side.

There is an infinite variety in the grains of this wood, and the illustrations given herewith must not be considered as the only types of the wood.

The primary object of these illustrations and, in fact, of all succeeding ones, is to exhibit the work done by the processes that are described, and the effort has been made to give the characteristic features of each wood in the limited space covered by the illustrations.

CHAPTER XV

SILVER MAPLE

Ground-color. — White lead with a touch of burnt sienna.

Graining Color. — Drop-black or ivory black, adding a little burnt sienna for overgraining.

Tools. — Same as for bird's-eye maple.

This wood is a gray variety of maple, and is often stained to produce a silvery effect.

The mottling and graining are practically identical with the methods used for bird's-eye maple. There are fewer bird's-eyes in the wood than in bird's-eye maple. The mottling should be made clean and distinct, and when properly overgrained, it makes a very effective piece of decorative work. It is, however, seldom used, and is less in favor than bird's-eye maple. In some cases it is used on the stiles of doors where bird's-eye maple is used on the panels.

CHAPTER XVI

WHITEWOOD

Ground-color. — White lead, yellow ochre, and raw umber.

Graining Color. — Raw umber, raw sienna, burnt sienna, black.

Tools. — Rubbing-in brush, combs, rags, pencil fitch, sash tools, and overgrainers.

This wood is known through the south and southwest portion of the United States, of which it is a native, as yellow poplar, and in color varies from a very light creamy shade to a dull grayish brown, in some cases nearly as dark as walnut. It has often a fine-grained heart, but is most frequently marked with reddish brown to gray streaks which run longitudinally with the grain. It is rarely found with any mottled effect, but there are exceptions. One variety is called blistered whitewood, for the reason that its markings resemble blisters in their shape. It is seldom necessary to imitate this variety.

This wood can be represented in either oil or water color, but it is most frequently done in oil. Its grains are very simple, and are seldom prominent in the heart grains, so they can be represented quite successfully with a bristle fitch tool or the small fresco liner.

In mixing the graining color use mostly raw umber for the base of the color. Tone this up with raw sienna, and after thinning apply a very thin wash of color to the work; then streak the plain portions with a thin wash of the graining color which has been slightly darkened by the addition of raw umber and a little burnt sienna. If necessary, for very dark veins a trifle of black may be added to the color. Draw the rubbing-in brush evenly over the work, using considerable pressure on the sides of the brush. This will leave a faint grain, and the dark veins can be made on the background and blended across the grain with the rubbing-in brush. Avoid strong contrasts unless the surrounding woodwork has examples of that kind.

The heart grains can be put in with some of the rubbing-in color which has been slightly darkened by the addition of a little raw umber or a touch of black, using the flat fresco bristle liner to apply the color and blending lengthwise. They nearly always appear as dark lines on a lighter background. Should they appear light, they can be represented by making a very fine outline grain with a pointed piece of wood covered with two layers of thin cotton cloth. Blend slightly lengthwise of the grain.

Whitewood seldom requires overgraining, as the divisions can be cleanly made and lighter and darker boards represented by varying the amount of color applied to the work in the first graining.

It is seldom necessary to grain it, as the original is now one of our cheapest woods. It is an excellent wood for panels, as it contains no sap; therefore it is unnecessary to apply shellac before it is painted.

CHAPTER XVII

SATINWOOD

Ground-color. — White lead, tinted with raw sienna.

Graining Color. — Raw sienna, burnt sienna, Vandyke brown.

Tools. — Sponge, rubbing-in brush, mottler, blender, and overgrainer.

GRAINING, ANCIENT AND MODERN

This wood is a native of India and belongs to the cedar species. Its figure is similar to that of mahogany, and especially so in the crotch of the tree. In the olden times these crotches were cut into thin veneers and used for panels on the finest of light cabinet work.

There is a variety of maple very similar in grain to the mottled satinwood, but the crotch or feathered effect is lacking.

This wood can be represented most effectively in water colors. Mix a thin wash composed of seven-eighths of raw sienna to one-eighth of Vandyke brown. Have some of the thick color of either shade on a palette or plate; rub in lightly with the graining color mixed with one-third beer to two-thirds water. If the panel is to show the feathered effect, a sash tool charged with a mixture of raw sienna and a very little Vandyke brown is used to lay in the darker portions of the grain on the color already rubbed in. The mottler is then used to take out the lights by removing portions of the dark color and working out the general plan of the feather, breaking up the continuous lines laid on by the sash tool. The badger blender is then used to soften the lines, blending crosswise. This work must be done with rapidity, as the water color soon dries. A little soap added to the color will make it dry more slowly. If the panel dries before the desired background effect is obtained, it may be wet with a clean mottler or overgrainer, using clear water not too profusely applied. The mottler, or cut tool, can then be used again to take out the high lights and produce the mottled effect, and the blender again used, blending across the panel.

Care must be taken to get but little color on the work. The most common fault of grainers who represent maple or satinwood is that they get the color too dark because they apply the color too freely.

When the background is dry, it should be overgrained, using the short-haired overgrainer and adding a little burnt sienna to the graining color. The overgrainer is charged rather sparingly with this color and the color applied over the feathered pattern, which should be dry when overgrained. The badger blender is again used to draw the color (applied by the overgrainer) to a sharp edge. Do not make the overgraining color too strong. If this is done, it tends to obscure the light and shade of the work. It is well to use bone combs of different sizes for separating the bristles of the short-haired overgrainer. The piped camel's-hair overgrainer can occasionally be used to advantage, especially on the feathered panels.

The mottled form of satinwood is far more frequently seen than is the crotch or feathered variety. This is represented by first rubbing-in the graining color and with the sash tool applying streaks of color slightly darker than the rubbing-in color. Then take the mottler, or cut tool, and separate the streaks or veins of darker color into short, darkened patches or groups and blend quickly with the badger blender. Care must be taken to keep the mottler, or cut tool, perfectly clean, rinsing it occasionally in clean water and wiping it on a clean cloth or wash leather.

When the mottling is dry, it should be overgrained, adding a little burnt sienna to the color, but being careful not to get the overgraining too strong, as one of the chief charms of the wood is its delicacy of figure and its suggestion of the folds of satin in its mottled markings.

Very pale varnish should be used to finish the wood. It is sometimes necessary to apply a thin coat of varnish before overgraining. In that case the general character of the work can be strengthened without danger of spoiling the work, which danger is always present when the panel is only wet over in water.

CHAPTER XVIII

WHITE MAHOGANY

Ground-color. — White lead, raw sienna.

Graining Color. — Raw sienna, burnt umber, drop-black, crimson lake.

Tools. — Sponges, rubbing-in brush, sash tool, mottlers, blender, and overgrainers.

This wood is grown on the east and west coast of Mexico. Its grain is very delicate and some portions are similar to satinwood, but the mottlings are seldom as strong as those of satinwood. Its color is also much lighter and the general character of the figure is more subdued and quiet in tone. It possesses unmistakable signs and veins of the true mahogany and is similar in general character to the grain of mottled mahogany, but exceedingly light in color.

I cannot recall having ever seen a feather or crotch of white mahogany, but there are doubtless such figures in the wood. All the panelled work that has come under my observation has had the mottled effects in the grain. There is very little difference in color between the long rails and the cross stiles, hence it would be unwise to make any strong contrasts in the graining.

This wood can be done most effectively in water color. Use raw sienna for the base of color; add a trifle of drop-black and a touch of crimson lake. Rub in with this color thinned with one-third stale beer to two-thirds clean water. Darken the color slightly by dipping the sash tool lightly in some of the very thin drop-black with just a touch of crimson lake added. Lay out dark veins in the general direction you wish them to appear. Then use the mottler to break them into short sections, or the back of the fingers may be used, as in maple; a small sponge may be used for this purpose.

If it is desired to have the work very light, the black may be omitted and a touch of burnt sienna or Vandyke brown substituted; or the mottling may be done directly on the rubbed-in work without the addition of any darker color. Before the color dries it should be softened carefully with the badger blender very lightly, always in one direction and across the grain of the wood.

A careful inspection of this wood will reveal a fine porous effect, which can best be obtained by carefully stippling the work. This can be done on the work when dry after it has had a thin coat of varnish. Or, it may be stippled on the ground-color and bound with a thin coat of varnish before the figured work is put in.

For stippling color use a very thin wash of raw sienna slightly deepened with Vandyke brown or black. If the work is quickly done, there is a chance to put in some stippling with the ends of the blender before the mottling is dry; or a round blender may be dipped in some darkened color and carefully used to make the porous effect.

The overgrainer must be used on the dry mottling, taking care to have the color very thin. Darken the overgraining color with a very little burnt sienna, just enough to make the overgrain perceptible without appearing too strong; blend lightly across the work with the badger blender.

CHAPTER XIX

ASH

Ground-color. — White lead, yellow ochre, and a touch of raw umber.

Graining Color. — Raw sienna, raw umber, and a touch of drop-black.

Tools for Oil Color. — Rubbing-in brush, sash tool, fitches, fresco liner, combs, rags, etc.

Tools for Water Color. — Sponge, rubbing-in brush, stippler, sash tool, fitches, fresco liner, overgrainers, crayons, etc.

This wood is distributed very generously in all temperate climates and is an excellent wood for timber of all sorts. Some of its varieties are very beautiful. It is one of the lightest woods in color, but is classed as a hard wood; in some varieties the grain is very dense, and in weight it is almost as heavy as oak.

The grains of ash, as a rule, are less intricate than those of any of the light hard woods, but occasionally we find an exceedingly complicated figure. The grains of Hungarian ash are very complex. Burl ash, which is an excrescence cut from the side of the tree, is a mass of fine grains, in some cases similar to the grains of bird's-eye maple. This sort of wood is generally sawed or cut into thin veneer and is used only for panels.

The heart grains of ash are less angular than almost any other of the light hard woods. The points of the heart grains are nearly always rounded and seldom run to a sharp point; this is the chief characteristic of this wood and it can readily be distinguished from chestnut (whose grain it often resembles) by this feature.

Previous to the advent of quartered oak for interior finish in Massachusetts, ash was the wood most in favor for a light wood. It had supplanted chestnut for this purpose. The latter was the only light hard wood in general use for interior finish in eastern Massachusetts when the writer went to serve his time with his father in 1872.

Since the cost of quartered oak has become so great, we see the ash again coming to the front. Being a softer wood, it is more easily worked than oak, and some people think that all light hard wood is oak, and often the cheap builder who wishes to sell his new house does not scruple to tell intending purchasers that it is finished in oak, when ash has been used.

In mixing the graining color use about equal portions of raw sienna and raw umber, adding a little drop-black. If the shade of the color is too yellow, use less raw sienna and more raw umber. For some shades of ash raw umber alone makes a good graining color. For oil color, thin with the prepared thinners previously described. If the color is not sufficiently thick, add some bolted whiting and stir it well into the graining color. The use of whiting is not recommended in oil colors where it can be avoided, as the traces of lime in the whiting attack the iron compounds in the umbers and siennas and make them bleach out or fade.

In graining this wood in oil color, first lightly sandpaper the groundwork, which should be about an egg-shell gloss, and dust off clean. Then apply the color evenly to a door or to several, as the color works better for being slightly set. Then use the rubber comb, not covered with a rag, or the steel combs covered with a thin piece of cotton rag, and make the grains, in the portions of the work intended to be left, in plain grains.

A little color slightly darkened with a touch of black may be used to put in some darker veins among the combed work and the rubbing-in brush used for a blender to soften the lines. A little of this work used with discretion serves to relieve the plainness of the combed work and presents an appearance similar to the natural wood without the expenditure of any great effort.

The heart grains are always placed in the panels, and in some cases, where good natural wood is to be matched, they are to be placed on nearly every board. They can be represented by wiping out the oil color or by being pencilled on the color with a fitch tool or liner, or crayons may be used for this purpose.

To wipe out the hearts, we use a soft piece of cotton rag — a piece of an old sheet is best for this purpose; fold the rag twice and place the thumb on the folded edge,

GRAINING, ANCIENT AND MODERN 49

allowing the centre of the nail to be covered by the rag, but the inside edge of the nail is not covered. Draw the loose ends of the rag with the other hand, and using that hand to steady the other, outline the work with the thumb, making sweeps of the arm. The outside edge of the wiped outline should be cleanly cut. The inside edge may later be softened by using the rag on the ball of the thumb. Draw the outlines carefully and carry out the lines on the outside edges of the heart grains with a two-inch comb of rubber covered with a thin piece of rag so that the panel will have the same appearance all over. A common fault in doing this work is that the places where the comb has been used can readily be distinguished from the work done by the rag on the thumb.

When the outline of the grain is completed, the rag can be placed over the tip or on the ball of the thumb, and the inside edges of all the outline softened by removing portions of the color. Look carefully at the natural wood and notice how to do this. Do not remove too much color.

When the work has been cleanly outlined and softened, a medium steel comb may be used to slightly serrate the lines of combing on the sides of the heart grains, but this is not always necessary. Then take the rubbing-in brush and blend the work lightly, always drawing the brush toward the points of the heart grains, which will slightly sharpen them and give them a more woody appearance; when the work is dry, it may be overgrained.

The heart grains of ash may also be pencilled in oil color, or the pencil can be used to interline the work wiped out by the rag, and when properly blended, it greatly improves the appearance of the wiped-out heart grains.

The flat fresco bristle liner shown in cut of tools is an excellent tool for this purpose or for making the heart grains of any wood. The pencilling color should be slightly darker than the graining color with which the work was rubbed in. A soft piece of rag may be used to wipe off the color previous to putting in the grains, or the color may be left as rubbed in. It should be slightly set before the pencilling color is applied, otherwise it is likely to run together too much. Draw carefully the outline of the heart grains. Hold a sash tool in the left hand and as each dip of color is taken in the fresco liner, rub it lightly on the sash tool so that the latter absorbs some of the color and leaves the liner sufficiently supplied with color without being too full after taking each dip.

When a panel or piece of work has been outlined with the liner, use the dry rubbing-in brush for a blender and blend lightly so that a sharp edge is formed on one side of the lines made by the liner. The blending is done toward the inside of the outlined heart grains, brushing the sharp edges of the color toward the outside of the heart grains. This will be found to resemble the grains most frequently seen in the natural wood. There are frequently exceptions to this rule, and careful attentions to the grains of the wood will show the student the best way to imitate it. Never allow the color to become so set that when the blending is done it will lift the color and show the ground-color too plainly. Such work is too scenic to be natural, and while admired by the amateur, is not so good an imitation of the wood as is more modest work.

Overgraining greatly improves the work. Take some of the rubbing-in color, thin it with spirits of turpentine, add a little black if necessary, and give a thin wash of color to the more prominent heart grains and wipe out any high lights. Refrain from attempting any knots either in the wiped-out work or in the overgraining. Knots are rarely seen in ash.

The work can be overgrained in water colors and, while it is a slower process on account of first having to dampen in the work with a sponge and a little whiting mixed with beer and water, it is most effective, and has the merit of keeping the oil color off the oil graining, so that a surplus

GRAINING, ANCIENT AND MODERN 51

of oil color is not applied under the varnish, which in time might cause the varnish to crack.

Aim to produce a woody appearance of the work as a whole. Do not overcrowd the doors with strong heart grains, and keep all joints and divisions cleanly cut.

Ash may be represented very successfully in water colors. In this case, little or no wiping-out is done, the heart grains being applied with the bristle liner. Thin the color with one part stale beer to two parts clean water. If the color crawls or creeps on account of lack of affinity with the groundwork, it may be remedied by rubbing the work with dampened whiting, or a cake of soap may be used, and the rubbing-in brush rubbed frequently on the soap. When the color is applied, it may be lightly stippled with the long-haired stippler, or if the work is very light, the stippling may be omitted. If stippling is done, the color should be very thin and the stippling finely done, not coarse.

The heart grains can then be put in with the bristle liner and the work blended with the badger blender. The short-haired overgrainer is used to carry out the lines on sides of panels. Where the liner has been used in the centre, they can also be used for the same purpose as the combs, and used in oil color. If the color is thickened, the rubber combs may be used with success. When the graining color is dry, it may be overgrained in thin oil colors. Crayons are sometimes used for making the heart grains, generally in water color.

HUNGARIAN ASH

Ground-color.—White lead, yellow ochre, chrome yellow.
Graining Color. — Raw sienna, raw umber.
Tools. — Similar to those for ash.

This wood is a native of southern Europe. It is more yellow in tone than light ash, and some varieties sold for Hungarian ash are undoubtedly American ash of a curly variety. Sometimes an old knurly ash log is cut into veneers by cutting around the circumference of the log, and the figure produced is similar in character to Hungarian ash, but is much bolder.

This wood can be imitated in either oil or water color. If done in water color, the work must first be mottled, using a thin wash of raw sienna and raw umber. Use the mottler or small fitch tool. Do not make the mottles too strong, as they can easily be darkened later on if necessary. Use a crayon to put in the grains on the mottling, or a camel's-hair pencil dipped in some darker color. The blender must be used to soften the lines made by the camel's-hair pencil. When dry, it may be overgrained in oil color and the mottling accented.

In graining Hungarian ash in oil color, the color may be applied in the usual manner, and the outline of the work wiped out with a rag. This is a tedious and intricate process and can be surpassed by work done by a camel's-hair pencil, or by the bristle lining fitch or a crayon pencil. In fact, the more I study the heart grains of any wood, the more I feel convinced that more successful work can be done by the application of color to a properly prepared foundation than by the removal of color from a similar foundation.

If the figure is to be applied with a pencil, first mottle the work, taking off the surplus color with a soft rag and blending lightly with the rubbing-in brush. This mottling can be allowed to dry if necessary or the grains can be pencilled in at once. The pencilling color should be slightly darker than the rubbing-in color, and it is well to thicken it with a little whiting or by the addition of fine, dry, raw sienna and raw umber. Pencil in the grains and blend lightly with the rubbing-in brush. A split steel comb may be used to break up the continuous lines and represent more faithfully the pores of the wood. Be sure and keep the teeth of the comb well cleaned. When this work has been allowed to dry, it may be overgrained in oil or in water color, but take pains not to get the work too dark, and endeavor to keep the color slightly different from that of the plain ash. The use of this wood is confined almost wholly to panelled work, so its imitations should only be found on similar work. Do not confound this wood with a species of curly ash which is similar in color to light ash but whose grains are different.

BURL ASH

Ground-color, graining color, and tools similar to ash.

Burl ash is an excrescence or abnormal growth which sometimes appears on the side of an ash tree. The late Professor Horsford of Harvard University, in his work on the Norsemen and their early occupancy of the North American continent, claims that one of their chief industries was the cutting of these burls from ash and oak trees and their sale in European countries, where the wood was used to make spoons and other articles of household utility.

The grain of burl ash is best represented in water colors. Sponge in the work and rub in the color, not too dark. Take a small sponge with medium-sized pores and dip it in some darkened graining color. Have a plate or palette and put the sponge on it to remove surplus color. Then apply the color to the panel evenly and endeavor to represent the little knots or dots that appear in the work. After going over the work, allow it to dry and then go over it a second time in a similar manner, using color darker than that used the first time.

If the panel is a large one, introduce a plainer grain toward the edge of panel and use the overgrainer to suggest the grains of the wood, merging the grains into the burl grains. When dry, rub the hand lightly over the work to remove any surplus color, and if necessary overgrain in oil color and slightly darken some of the clusters of dots or knots. There is very little light and shade in the average specimen of this wood.

Burl ash can be imitated in oil color by using a sponge in the manner suggested for water color. First dip the sponge in water and wring out dry, then take up the oil color on one side of the sponge only. When the work is finished, the sponge can readily be cleaned by the use of soap and hot water.

Burl ash panels, like those of Hungarian ash, should be used chiefly on interior work and then only in the panels of doors or wainscots. It is unwise to imitate them on outside work, as the real wood would rarely be used for that purpose.

CHAPTER XX

QUARTERED OAK

Ground-color. — For light oak: white lead, yellow ochre, and a touch of burnt umber. For dark oak: white lead, yellow ochre, raw sienna, burnt umber. By varying the quantities of these colors the ground-color can be made as dark as necessary.

Graining Color. — Raw sienna, Vandyke brown, drop-black, burnt umber.

Tools for Oil Color. — Rubbing-in brush, sash tool, fitch tools, bristle liner, combs, rags, etc.

Tools for Water Color. — Sponge, rubbing-in brush, sash tool, fitches, bristle liner, overgrainers, combs, blender, etc.

Quartered oak, or more properly quarter-sawed oak, is produced by sawing the wood parallel or nearly parallel to the medullary rays which radiate from the centre of the tree toward the bark. The log is usually halved or quartered, and boards sawn from the flat sides of each quarter until the section is cut away. The nearer the wood is sawn parallel to the medullary ray, the more prominent and eccentric appear the grains of the wood. Undoubtedly

GRAINING, ANCIENT AND MODERN 55

the locality and soil in which an oak tree grows affects in no small degree its markings. The white oak tree of New England and the Middle West when grown in favorable soil produces when properly sawn the boldest effects of quartered oak, many of the medullary rays appearing in the wood as large as the two forefingers of a man. The English oak tree when similarly sawn produces grains rarely one-fourth as prominent as those of American oak. The character of the figure is also different. The slowly developed English oak appears more dense in fibre and radically different both in heart grain and quartered effect from American oak.

In some varieties of oak the medullary rays are short and very thin, so that although the wood may be sawn parallel thereto, the effect produced is almost like that of pine wood sawn in a similar manner. Such wood is usually sawn to produce heart grains.

A section of a swamp white oak tree grown in South Carolina, measuring four feet in diameter, was exhibited in the Forestry Building at the World's Fair in Chicago in 1893, and again at St. Louis in 1904, and although sawed directly through the centre of the tree the medullary rays were scarcely discernible on the sides of the board where they should have appeared most prominently.

A friend of mine told me an amusing story of a capitalist friend who had recently purchased some valuable timber land in Tennessee. The man who sold the land said it was covered largely with oak trees, not the common oak but all quartered oak trees. (?)

Of all the woods that are represented by graining, oak, and especially quartered oak, is most frequently encountered. Go where you will in the civilized, world wherever the English language is spoken, and you will find but few localities, especially if adjacent to large cities or towns, where imitations of oak are not to be seen.

If samples of quartered oak, done by all the different

grainers of any large city, could be exhibited, they would often show what a wide variety of opinion existed among grainers as to how the wood should be imitated and what they considered its most beautiful figures. In many cities some able workman sets the style, and his apprentices and imitators are more likely to follow his style than to study and follow the grains of the wood.

Be original as far as possible. Copy no man's style or his work. Go to nature for your originals. Study closely the figure of the wood, use any or all methods to attain woody effects, but copy nature, not man.

Unless one is in love with his calling and cherishes its secrets and methods he rarely can accomplish the best results in his work.

A word might well be said at this point against the custom of staining wood and obscuring the beautiful figure of quartered oak. The so-called antique oak may have been growing as a white oak tree within twelve months of the time it is placed on the market as furniture. It is folly to stain it to a shade as dark as walnut and call it "antique." Its beauties are largely hidden, yet some of the furniture-makers claim that their treatment "develops and brings out the hidden beauties of the wood." As well might one say that the charms of a beautiful woman might be "developed" by the application of burnt cork or stain to her features.

The custom of staining light oak to an abnormal color has absolutely nothing to recommend it except the vagaries of a passing fad or fashion. It is similar in taste to that of the aborigines who paint their faces and bodies, hoping thereby to make themselves more handsome. Some furniture manufacturers do not hesitate to stain their oak furniture to a bilious green shade, unlike anything in nature, and the grainer is sometimes forced to try and represent it. This practice should be discountenanced by all intelligent people. If dark wood is required, there

GRAINING, ANCIENT AND MODERN

exist an abundance of varieties that are by nature dark and can be used for any purpose. But on general principles all natural wood should be finished as nearly as possible in its natural color. It will, as a rule, grow darker with age, but the man never lived who can improve on its natural color or figure. There is a growing sentiment favorable to a return to natural effects in furniture, and to a large extent the color of the furniture determines the color of the finish of the modern house.

WIPING OUT QUARTERED OAK IN OIL COLOR

When the work is properly rubbed in, it may be combed with a medium or fine rubber comb and overcombed with the split steel comb. When the proper effect of pores of the wood is obtained (which can readily be done as previously directed), a soft cotton rag is folded and the thumb placed in the rag, the loose ends of which are held by the left hand; the figure of the wood is then represented by removing portions of the combed work, and the intervening spaces can be softened by folding a small piece of rag and softening the combing with the folded edge of the cloth. Don't take off too much color. The edges of the wiped-out work may also be treated in this manner or the second joint of the forefinger may be drawn against the edge of each vein or flake. This will make the color appear slightly darker and when lightly blended a very woody effect is obtained. A small, flat, short-haired fitch used dry on the wet color will produce a similar effect. When dry, the work may be overgrained in either oil or water color. If oil color is used, some of the rubbing-in color thinned out with turpentine may be used. Rub in the panel and before the color sets comb with a medium-toothed rubber comb over the veins of oak and use the split comb to break the continuity of the lines made by the rubber comb; or a short-haired overgrainer may be used and the oil color applied directly to the panel without rubbing in, and the lines made by the overgrainer may be serrated and cut up by the split steel comb. This work must be done with great care to produce the effect of oak wood. Too much color will spoil the work and make it look "liney" and too regular for natural wood. Vary the shade of the overgraining color, also the width of the lines of color, and avoid overdoing it. Keep all joints and divisions cleanly cut, and in quartered oak add a few fine, dark veins to some of the plain combed work, not making the contrast too sharp.

GRAINING QUARTERED OAK

A fairly successful job of dark quartered oak may be done on a white pine or whitewood foundation without paint by first applying the quartered veins with white shellac, using a camel's-hair pencil or a short-bristle fitch. When dry, apply your oil graining color and try to comb it before the color strikes into the wood. If some extra megilp is used, this is possible. The portions of the work touched by the white shellac will appear light, and if necessary, they may be wiped clean with a rag, or instead of trying to comb the work it may be rubbed over with the rag and the surplus color removed, leaving the portions light where they have been covered with the white shellac. Such treatment cannot be very successful unless the wood is free from prominent heart grains, as they appear through the quartered veins and destroy the effect it is intended to produce. The check roller may be used to advantage on such work. A quick job can be done in this way, as a coat or two of shellac or varnish will finish the work in a very short time; yet the process is not recommended except in extraordinary cases.

We sometimes see an exemplification of this process on cheap oak furniture, where the wood has been sawn to produce the heart figure and the attempt is made to have it appear as quartered figure by the process described above. The dark stain, penetrating the pores of the wood, leaves the quartered figure represented by the shellac standing out boldly on a heart-grain background, or the same effect is produced by finishing the work light and after applying a dark stain wiping out the figure of quartered oak. In either case the deception is readily discerned.

Avoid the temptation to make bold figures all over the work. Repose and balance are essential factors in the appearance of any piece of figured work, whether it be the natural wood or an imitation.

We frequently see, in the natural wood, very bold effects; but the effect is not pleasing when these are combined in panelled work by a joiner who pays no attention to the figure of the work, but makes the boldest and most violent combinations of figure. While we must admire the beauty and eccentricity of a bold-veined panel of quartered oak, its beauty can be enhanced by keeping the surrounding stiles in finer grains and less bold than the panelled work; or if the panels are of fine-grained figure, the stiles may be made more bold; but always avoid the appearance of overcrowding the work.

The poet Oliver Wendell Holmes once wrote, "and since I never dare to be as funny as I can." It would be well for some professional grainers to take this advice to heart and not try to make their work too eccentric or funny, nor try to outdo themselves in the abundance and boldness of their work in a limited space. This is one of the most common faults of even the professional grainer and one of which he is often quite unconscious.

In these days when we see so much quartered oak in furniture and interior finish, it is unwise to mingle the heart grains among the quartered veins. If the work is to be quartered oak, see that no heart grains are introduced unless it be on the edge of a board, as is the case in the natural wood. There is such infinite variety in the grain of quartered oak that ample opportunity is afforded to show the skill of the workman without necessarily repeating figures.

FOURTEEN WAYS OF IMITATING QUARTERED OAK

1. Probably the oldest method of imitating quartered oak was by representing it in water color by first marking out, with a tallow candle on the painted groundwork, the figure intended to be shown on the work. Then rub in the work with water colors with the usual proportions of beer (one-third to one-half), and with the blender or with overgrainers make the effect of the pores of the wood. When the water color is dry, a soft dry cloth is rubbed over the work, which will remove the tallow and not disturb the water color which has adhered to all parts of the work not covered by the tallow. The work can then be overgrained if necessary.

2. Another method is to lay in the work in water color without making any strong longitudinal grains and allow it to dry. Then mark out the pattern of the quartered oak figure with a tallow candle and sponge off the surplus color. This will leave the veins dark on a light background, which is the reverse of the first process. The work, if carefully done, can again be gone over in water colors, and overgrained, but it is safer to use oil color and do the combing in the usual way. The color can then be wiped off the veins, and a margin of the groundwork can, if necessary, be shown, which will give the work a very effective appearance. Avoid strong contrasts of color in using this process.

3. The dark veins may also be represented by using a short-haired, flat bristle fitch or a camel's-hair pencil in water color. First dampen in the work by using one-half stale beer to one-half water. Add a little whiting if the color creeps or crawls. Use little or no color in this process. Then take a little dark color in the fitch tool or

pencil and apply the dark veins wherever desired. Use the blender very lightly, and when dry, overgrain in oil color not too dark. It will be unnecessary to wipe the color from the dark veins after the oil color is applied unless the graining color is very dark.

4. Another method to show dark veins on a light background is similar to number 3, but instead of the tallow candle a little beeswax is melted in turpentine and in linseed oil, and the water-color work is painted with this compound wherever the dark veins are to appear. A little dry color added to the beeswax shows more plainly where the brush touches the work. After sponging off the water color the surplus oil and beeswax may be removed with a dry cloth and the work overgrained in oil.

5. The dark veins may be painted in oil color, directly on the groundwork, and after they are thoroughly dry the work may be overgrained in water color or in oil color. The spaces between the veins can be more successfully treated at this time than if the work were all done in the wet color.

6. Another method to produce light veins on a dark background is to take some damar varnish, add to it a little lampblack, and paint on the work the figures it is desired to have appear in light color. When dry, overgrain in water color, and endeavor to produce the effect of the pores of the wood. When the water color has dried, dampen a rag with turpentine and wipe off the black veins. The turpentine will not remove the water-color graining, but leaves the light veins clear and distinct on a dark background. The work can then be overgrained in oil color.

7. Light veins may be produced on a dark background by rubbing in the work in water color, and when dry wiping out the figure with a wet chamois skin or with a wet rag, and when dry overgraining in oil colors.

8. Dark veins may be represented in the wet oil color

after the work has been combed and allowed to set a little while, by taking a little oil color in the bristle liner and applying it to the work, making the figures of the wood. The rubbing-in brush is then used as a blender and the work lightly blended in one direction. This will lift the graining color and show a rather light vein with a dark shadow on one side. Hence care must be taken not to lift the color too much, unless it is desired to take it all off and show the veins light. This can be done if the color is allowed to set and clear thinners be used for the pencilling color.

9. There is perhaps no better way to represent the dark veins of quartered oak than to paint them in oil on the dry combed work. Use color slightly darker than the graining color, and add a touch of burnt sienna, as these veins are frequently of a reddish tone. A square-edged, short-haired bristle liner makes the best tool for this purpose. The combing should not be too dark or too strong, merely a background for the color. A close examination of the dark veins in the real wood will reveal the fact that the background in such work, which we try to imitate by combing, is, as a rule, very subdued, and the dark veins are the most prominent in color of any of the grains; yet they should not be imitated too strong — better to have them too faint than too bold. The spaces between the dark veins can be glazed over with some thin graining color, and the effect of the natural wood is more frequently produced in this way.

10. Another method, and one rarely attempted, or indeed necessary to the skilful workman, is to prepare pieces of stout manila paper or tin-foil cut into the shape of the veins of quartered oak and attach them to the groundwork with weak paste. Grain the work, combing the parts covered with the paper or the tin-foil. When it is dry, remove these pieces from the work, and the figure will stand out boldly, especially if the graining color is rather

TOOLS

1. Stippler.
2. Large overgrainer.
3. Short-haired overgrainer.
4. Bone comb (for separating bristles of overgrainer).
5. Large mottler.
6. Small mottler.
7. Badger blender.
8. Piped overgrainer.

PLATE 2

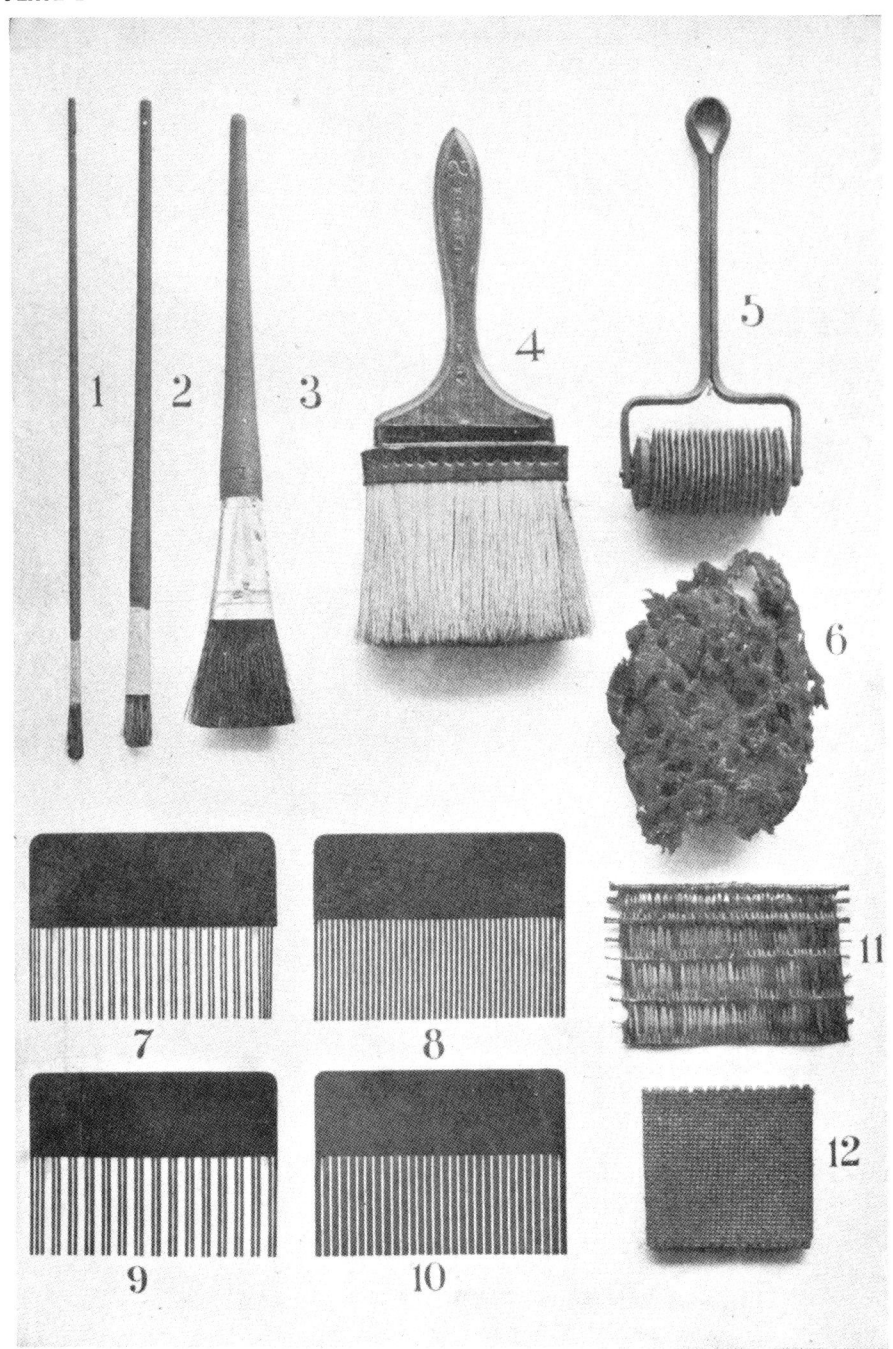

TOOLS

1. Flat fresco bristle liner.
2. Short-haired liner, or fitch tool.
3. Sash tool.
4. Rubbing-in brush.
5. Check roller.
6. Sponge.
7. Fine split steel comb.
8. Medium steel comb.
9. Medium split steel comb.
10. Coarse steel comb.
11. Selvedge of straw matting.
12. Rubber comb.

PLATE 3

TABLE TOP — GRAINED TO REPRESENT INLAID WOODS
Size of top. 28 × 42 inches. Varieties of wood represented, 14. Number of pieces, 12,426.

PLATE 4

TABLE TOP—GRAINED TO REPRESENT INLAID WOODS
Size of top, 22 x 32 inches. Varieties of wood represented, 12. Number of pieces, 5485.

PLATE 5

"JERSEY OAK"
Done in Cambridge, Mass., 1845.

PLATE 6

CURLY MAPLE — MOTTLED
To overgrain

CURLY MAPLE – OVERGRAINED

PLATE 8

BIRD'S-EYE MAPLE
First stage

PLATE 9

BIRD'S-EYE MAPLE — OVERGRAINED

PLATE 10

BIRD'S-EYE MAPLE — FINISHED

PLATE 11

WHITEWOOD

PLATE 12

SATINWOOD — FEATHER PANEL

PLATE 13

WHITE MAHOGANY

PLATE 15

LIGHT ASH — WIPED OUT AND PENCILLED

PLATE 15 A

LIGHT ASH — WIPED OUT AND PENCILLED

ASH – HUNGARIAN ASH PANEL

PLATE 17

ASH - BURL PANEL

PLATE 18

DARK ASH

PLATE 20

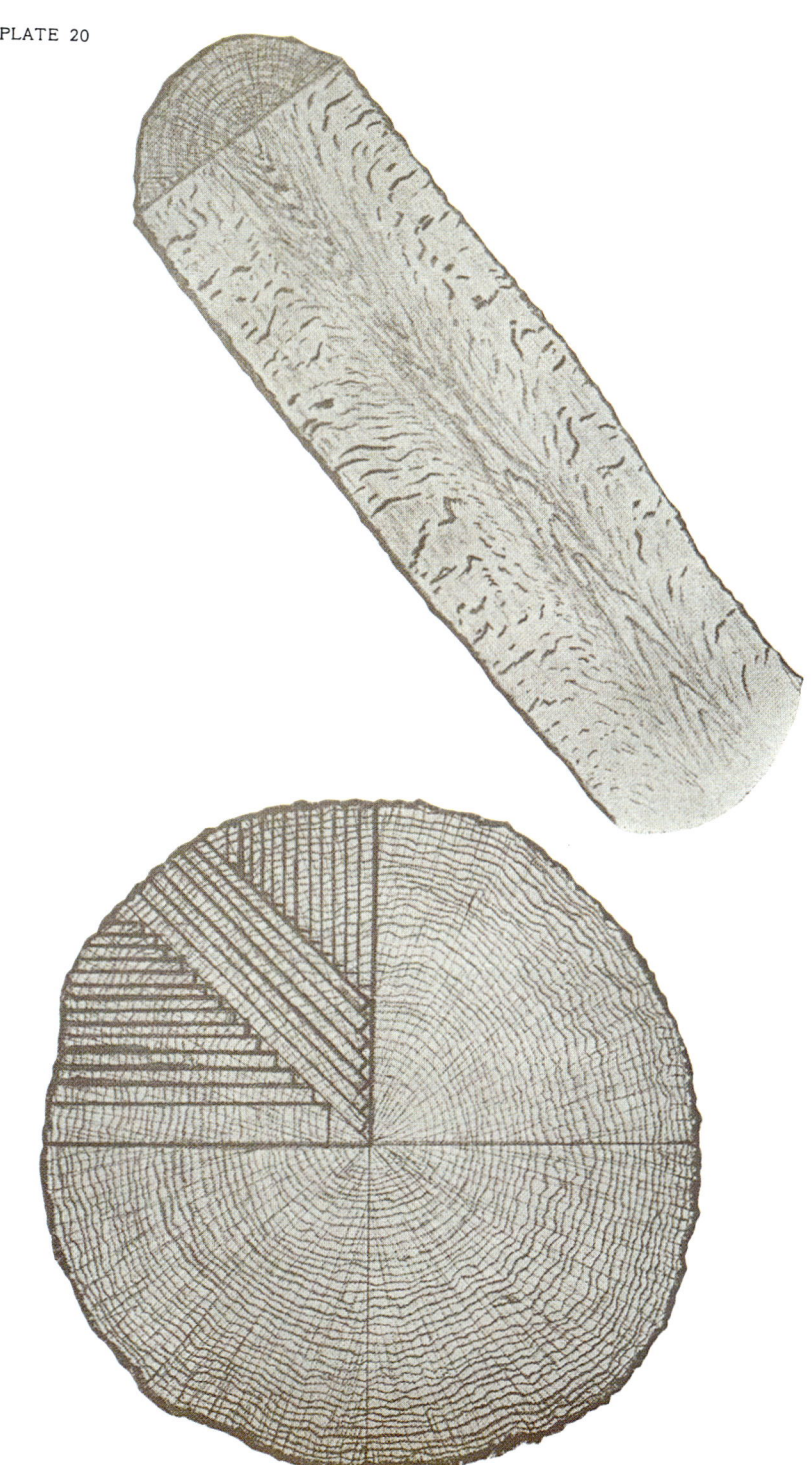

1. HEART OF OAK LOG SHOWING QUARTERED OAK ON EDGES
2. END OF OAK LOG SHOWING HOW QUARTERED OAK IS SAWED

OAK — COMBED
Ready for quartered veins

LIGHT QUARTERED OAK — OVERGRAINED

PLATE 23

LIGHT QUARTERED OAK — OVERGRAINED

QUARTERED OAK — DARK PANEL

QUARTERED OAK

QUARTERED OAK (IN WATER COLOR)

LIGHT QUARTERED OAK

PLATE 28

LIGHT QUARTERED OAK

PLATE 29

ENGLISH QUARTERED OAK — POLLARD OAK PANEL

ENGLISH OAK — ROOT OF OAK PANEL

PLATE 30

DARK QUARTERED OAK

PLATE 31

DARK QUARTERED OAK

PLATE 32

DARK QUARTERED OAK

HEART OF OAK – LIGHT

PLATE 34

DARK HEART OF OAK

CHESTNUT

PLATE 41

OREGON CEDAR

PLATE 42

YELLOW PINE

PLATE 43

PITCH PINE
or Hard Pine

PLATE 39

CYPRESS

PLATE 14

QUARTERED SYCAMORE

CHERRY — MOTTLED
Ready to overgrain

PLATE 36

CHERRY — OVERGRAINED

PLATE 37

CHERRY — OVERGRAINED

PLATE 40

CURLY BIRCH

STIPPLING FOR WALNUT OR MAHOGANY

PLATE 49

BLACK WALNUT — OVERGRAINED

PLATE 49A

BLACK WALNUT

PLATE 50

CURLY WALNUT

PLATE 51

BLACK WALNUT — BURL PANEL

PLATE 52

ITALIAN WALNUT

CIRCASSIAN WALNUT

MAHOGANY — MOTTLED PANEL

MAHOGANY — FIGURED

PLATE 46

MAHOGANY — FEATHER PANEL

MAHOGANY — FEATHERED PANEL

PLATE 47

TEAK

PLATE 54

ROSEWOOD
First stage

ROSEWOOD — OVERGRAINED

PLATE 55A

ROSEWOOD

PLATE 56

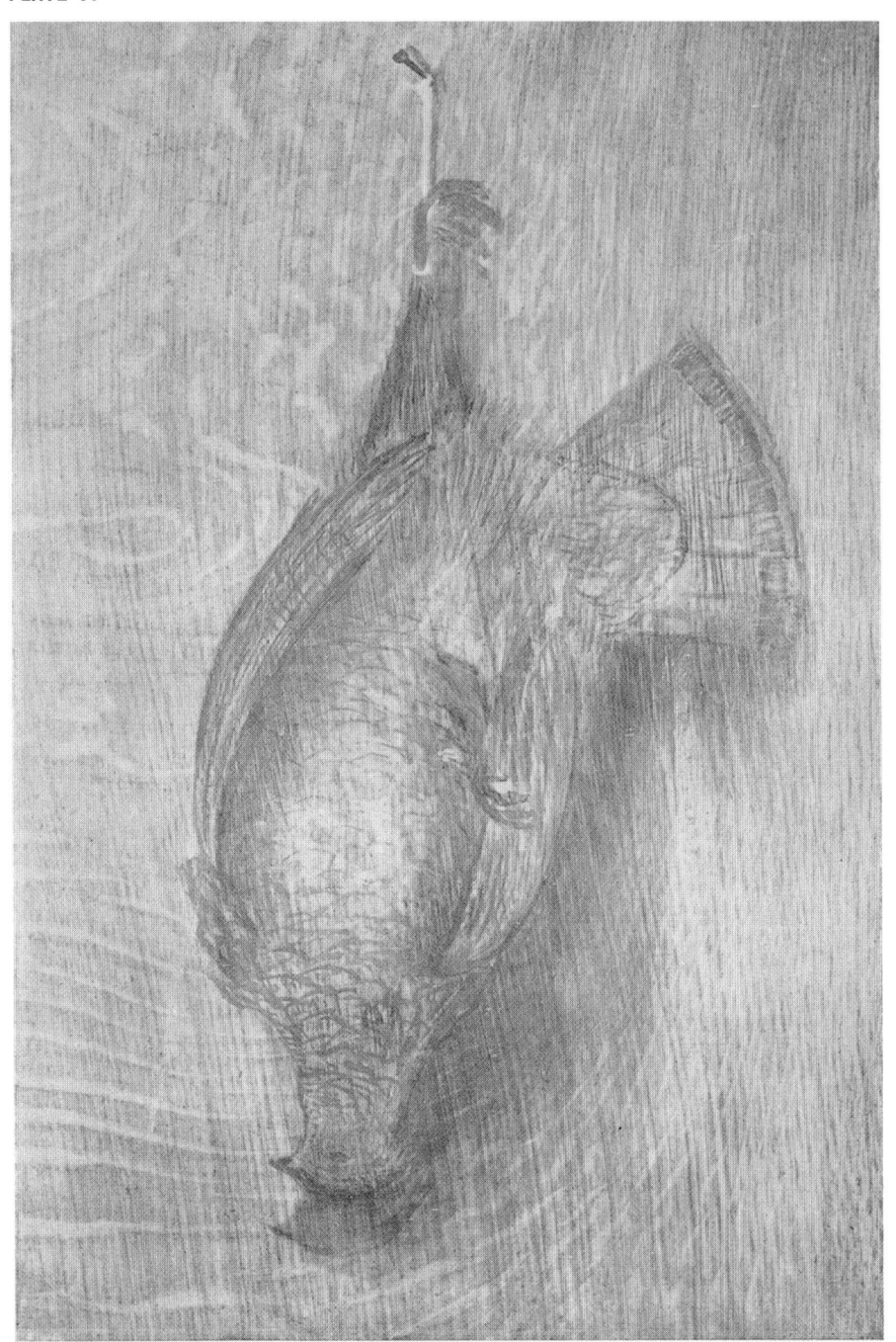

IMITATION OF CARVING

PLATE 38

BUTTERNUT

GUM WOOD

PLATE 58

OREGON OR DOUGLAS FIR

dark. This work should be overgrained to produce a woody effect.

11. Quartered oak may be represented on any light wood that has a clear grain, and is free from strong heart grains, by the use of white shellac, which is used to make the figure of the veins, applying the shellac with a bristle liner directly to the smooth wood, without previous preparation. When the figure has been outlined with the shellac and allowed to dry, which will be a few moments after being applied, the oil graining color, which has been mixed rather oily, and with an extra portion of megilp, is immediately applied to the work, and the rubber and steel combs are used to represent the pores of the wood. The graining color can then be easily wiped off the portions of the work previously coated with the white shellac.

The effect is fairly good if the wood is clear grained and transparent, but there is generally something in the figure of the soft wood which indicates that the work is counterfeit.

In some cases the work is not combed at all, but the color is allowed to sink into the wood, and the color on the face of the shellac is rubbed off with a soft cloth. In exceptional cases this work might pass for oak. It would be superior to the simple staining of the wood to an oak color, but it could not compare with work done on a painted foundation.

12. When the grain of the wood is clear and free from knots or sappy places, it is possible to do an excellent job of quartered oak by giving the work two coats of white shellac, and using that for a groundwork, graining in oil in the usual way. There is a transparency in such work that is lacking in any work done on a painted surface, but the wood must be clear and free from strong grains of any kind or the grain of the wood will spoil any attempt to introduce quartered oak veins into it.

13. Another method of imitating quartered oak is the

so-called spirit graining, a method seldom resorted to by the trade grainer.

Mix up bolted whiting with turpentine to a stiff paste; add raw sienna and burnt umber to make the desired shade of color; add some good japan dryer and a small amount of linseed oil; thin the color with turpentine to a working consistency. Strain the color and rub in a small piece of the work at a time and comb it immediately. Allow it to get well set and then use a small square-edged fitch tool or liner which has been dipped in a mixture of clear turpentine and raw or burnt sienna, and with this brush apply the markings or veins of quartered oak over the work previously combed. When the turpentine has softened the spirit color, which will be almost immediately, it can be rubbed off clean with a soft rag, leaving the figure light on a dark background. Diluted washing soda may be used for the pencilling color instead of turpentine, but there is danger of spoiling the work unless it is used very carefully. When dry, this work may be overgrained in water color.

14. Lastly, the method followed by probably the greater number of grainers who do quartered oak in oil color is to wipe out the veins with a soft cotton rag folded over the thumb nail, or with a veining horn, which is held in the palm of the hand and lays against the under side of the thumb, projecting slightly beyond the thumb nail and covered with a fold or two of the rag. Some of the old grainers use a piece of soft rubber for this purpose, but it is difficult to prevent showing a fat edge on the work where the uncovered rubber has been used. In some portions of the world grainers fold the rag into a sort of tape and put in the work with the rag drawn over the end of the forefinger.

CHAPTER XXI

ENGLISH OAK

Ground-color. — White lead, yellow ochre, raw sienna, burnt sienna.

Graining Color. — Raw and burnt sienna, burnt umber, Vandyke brown.

Tools for Oil Color. — Flat brush, sash tool, fitches, bristle liner, combs, rags, crayons.

Tools for Water Color. — Sponge, rubbing-in brush, sash tool, fitches, bristle liner, overgrainers, badger blender.

This wood is the native British oak, grown in Great Britain or Ireland. Its growth appears to be much slower than that of any of its American cousins. Its grains are often quite intricate and eccentric. The quartered oak sawn from English oak trees differs very much in figure from American quartered oak. As a rule, the grains are very much smaller and more eccentric. They are also interspersed with dark streaks, which gives the work an odd appearance.

The work is usually done in oil color, mixing a little raw sienna with Vandyke brown and then using the regular thinners to the desired shade. The combing should be more wavy than that intended to represent the grain of American oak and should be well cut up with the split steel comb. Next, wipe out the fine quartered veins with a rag and blend lightly. When dry, overgrain in water color and apply the streaks of short, dark grains with the lining pencil; blend at once with the badger blender. When dry, overgrain in the usual manner with a thin wash of oil color.

POLLARD OAK

Ground-color. — White lead, yellow ochre, raw sienna, burnt umber.

Graining Color. — Raw sienna, burnt umber, Vandyke brown, drop-black.

Tools for Oil Color. — Rubbing-in brush, sash tool, fitches, bristle liner, combs, etc.

Tools for Water Color. — Sponge, rubbing-in brush, sash tool, fitches, bristle liner, overgrainer, blender, crayons, etc.

This wood is the product of intelligent cultivation of the oak tree in European countries. The small limbs that extend outward are polled or cut into rounded heads close to the tree. As the tree grows, these polls or heads are included in its growth, and when the tree is cut for timber, these heads are cut through by the saw and the result is a very beautiful figure of knotted oak.

Pollard oak may be imitated in either oil or water color. The former method is that most commonly used. One of the best jobs I have ever seen was done in water color. It is most frequently represented by doing the first work in oil color and overgraining in water color. The figure of the wood consists of groups of dotted knots of greater or less diameter and masses of light and shade with both quartered figure and heart grains interspersed among them. To make these groups in oil color, darken some color with burnt umber and a touch of Vandyke brown, and apply it to the work in irregular patches with a sash tool or fitch tool. Then take a soft rag and work out the darker portions to resemble groups of knots. Remove some of the color from between the groups of knots and comb with the rubber and steel combs. Heart grains of the finer figures of quartered oak may be introduced among this work.

When the oil color is dry, the work should be overgrained in water color and the knots worked up to show the light and shade. Some of the lighter portions of the combed work may be covered with the water color, and when dry, the figure of quartered oak may be represented by using a wet chamois skin, a wet rag, or a slice of raw potato, either of which will remove the water color from the work. The crayon pencil encased in wood is an excellent help in doing this wood, and it may be used in either oil or water color.

ROOT OF OAK

Same colors and tools as for Pollard oak.

This wood presents more eccentricities of figure than Pollard oak. It is seldom imitated, but its grain is distinctly different from that of Pollard oak. There is in this wood a much larger element of twisted and curved lines than those seen in Pollard oak. The light and shade is also more sharply defined. It can be well represented in either oil or water color in the same general manner as in the process describing the imitation of Pollard oak.

CHAPTER XXII

HEART, OR SAP, OAK

HEART, or sap, oak is the grain as it appears when sawn the length of the log from bark to bark. As the centre of the log is approached the quartered grains begin to appear toward the bark of the tree, disappearing as they approach the heart or centre of the tree.

In ordinary white oak, and in much of the red oak, these quartered grains are very pronounced when the log is sawn parallel to what botanists call the medullary rays. These are the hard, bright flakes that appear most plainly in the end of a piece of oak, and which always radiate from the centre of the tree toward the bark. In most of the oaks used for timber, the log, when cut in this manner, produces the most beautiful effects in quartered oak. There are, however, some kinds of oak wood in which the medullary rays are so short that, although the tree may be sawn perfectly parallel to the medullary rays, the figure produced is so fine as to be almost invisible except at very close inspection. Such oak, when used for interior finish or furniture, is usually sawn to produce the heart grains, which are often very peculiar and seldom as graceful as those of white or red oak.

While it must be conceded that the finest work of art is inferior to the works of nature, yet in imitating the grains of wood it is unwise to spend time and labor in representing the inferior patterns of the grains to the neglect of the superior ones. After we have done our best we are often far enough away from our copy, yet we ought at least to have an ideal, and that ideal should be to faithfully represent the grains of the particular wood we are imitating and endeavor to produce the effect of the more beautiful figures of that wood; not entirely ignoring the plainer grains, but whenever the finer figures are called for in the work, we should be able to reproduce them. This means that all grainers should not only start with well-defined ideas of the grains of the different woods ordinarily used in interior finish, but they should also possess panels of these woods and constantly study them. Keep them in view. Let them not be relegated to some obscure corner nor hung close to the ceiling of some smoky old shop or office. Lose no opportunity to add to their number whenever you can do so. They need not necessarily be all of one size, but secure them as large as you can.

It is difficult to remove our first impressions, and if these impressions can be made to conform to the grain of the natural wood rather than to somebody's idea of the

GRAINING, ANCIENT AND MODERN 69

wood, we have at least the satisfaction of having started right, and are more likely to stick to nature for our ideals than to the work of the most skilful grainer, and to make all the criticisms of the work of ourselves and others from the standpoint of the natural wood rather than from any technical excellence of the work.

WIPING OUT HEART GRAINS OF OAK

There are a variety of methods of representing the grains of this wood. It is probably represented in oil color more frequently than in water color, and an excellent representation can be made by either method or by graining in oil color and overgraining in water color, or *vice versa*.

For some varieties of quartered oak it is difficult to excel work done with a crayon if rightly used.

In wiping out the heart grains in oil color the same general method is used as for wiping out the hearts of ash as described in a previous chapter.

The rag is folded and held over the thumb nail and the grains are outlined by removing the graining color from the ground color. The heart grains of oak are, as a rule, serrated and less rotund than those of almost any other wood. They also vary from very coarse to very fine and are often found taking an eccentric formation on either side of the main heart. Often there are small knots in the work, but as a rule these appear on the outside edges of the board.

It would be impossible to fully describe all that can be done with a rag and comb in wiping out the heart grains of oak. Nothing but diligent practice and careful observation of the real wood will help the learner to become proficient in this method.

The work should be well outlined with a clean-cut outer edge. When the color sets slightly, the inside edge of the outline can be softened with the rag folded or by covering the point of the thumb with the rag. Do not remove too much color. When the lines are carried out on the sides of the work with a rubber comb, the split steel comb may be used to further serrate the outer edges of the heart grains, or the blending comb previously described may be used.

Blend the work lightly with the rubbing-in brush, and when the work is dry overgrain it to put in any lights and shades. If necessary, use a camel's-hair pencil to make the fine checks or medullary rays on the work. The check roller can also be used for this purpose.

Another method of imitating the heart grains is to comb the work with a coarse steel comb, overcomb it with the medium split steel comb, and put in the heart grains with the bristle liner, using some of the rubbing-in color slightly darkened with dry burnt umber. Blend immediately and draw the color to a dark edge on the outside of the figure. If the lines look too continuous, the split steel comb may again be used to cut up the lines to resemble the pores of the wood.

Heart grains of oak, especially of some of the western oak, may be well imitated in water color; as a rule an undercoat of faint stippling is necessary. Use one-third beer to two-thirds water and a little burnt umber. When this has dried, the heart grains may be put in with the bristle liner and carefully blended with the badger blender. Care must be taken not to work up the underneath color. The work must be done expeditiously.

The grains are sometimes put in with oil color on the water-color stippling. Nothing but continued experiments will enable the learner to discover the method best suited to his taste or that in his opinion appears to represent the wood more closely.

A thin wash of overgraining color should always be applied over the heart grains to produce the most woody effects.

A very good imitation of the dark heart of oak is made

by using crayons in oil color. Rub in the work with but little color in the thinners and make the heart grains with a crayon. Cut up the lines with a split steel comb and blend lightly with the rubbing-in brush. When dry, the work should be overgrained in either oil or water colors.

Crayons can also be used dry on the stippled background in water colors. After the work is outlined, fill a medium-sized overgrainer with clean water and draw it lightly over the work toward the points of the hearts; this will wet the crayon lines. Blend immediately with the badger blender; this will draw the wet crayon lines to a sharp, dark edge on one side of the work. Steel combs may be used, if necessary, to break up the lines.

CHAPTER XXIII

CHESTNUT

Ground-color. — White lead, yellow ochre, raw sienna, burnt umber.

Graining Color. — Raw sienna, burnt sienna, Vandyke brown.

Tools for Oil Color. — Rubbing-in brush, sash tool, fitch tools, bristle liner, combs, rags, etc.

This wood is a native of nearly all temperate climates and makes a good wood for timber. It has a very coarse grain and is very porous, hence it is difficult to keep it properly filled so that the weather will not affect it.

It may be represented in either oil or in water color. If the latter, the work should first be finely stippled and the heart grains put in with the bristle liner and immediately blended with the badger blender. The plain portions of the wood may be represented by using the overgrainer.

For oil graining the color is mixed of two-thirds raw sienna to one-third Vandyke brown, adding a little burnt sienna if necessary. Thin with the regular thinners, and when the color has been properly applied and allowed to set for a little while, the heart grains can be wiped out with a rag in the manner directed for ash. Do not wipe off too much color. To make the fine secondary grain that usually is seen in this wood, use a thin piece of wood sharpened to a point, blend in lightly with the rubbing-in brush. Use the combs as directed for ash. There is very little fine combing seen in chestnut.

CHAPTER XXIV

WHITE OREGON CEDAR

Ground-color. — White lead, yellow ochre, venetian red. Make a warm shade, not too much lead.

Graining Color. — Raw sienna, burnt sienna, rose pink or crimson lake, drop-black.

Tools for Water Color. — Sponge, rags, brushes, fitches, overgrainer, mottler, blender.

Tools for Oil Color. — Usual brushes and combs.

This wood, which grows on the northern Pacific coast, is becoming more common in the eastern part of the United States, owing to the scarcity of native white pine of good quality. It is doubtful if the wood is as serviceable as white pine, it being very soft, exceedingly light in weight, and readily takes a dent or a bruise. It also splits easily if not carefully nailed. It is, however, one of the most durable of woods.

A photograph is exhibited by a Lumberman's Association of California, showing a man at work cutting out shingle sections of sound lumber from a fallen log of cedar; arched over the log are three growing cedar trees, each estimated to be fifteen hundred years old, showing that

GRAINING, ANCIENT AND MODERN 73

although the log had lain all those years it had not rotted, but remained sound timber.

For water color, use beer one part, water three parts; dampen with a sponge, rub in with brush dipped in a little of each of the colors named, lay out for light and shade and general direction of grain; blend, and when dry put in dark figure of heart grains, which almost always appear dark on a lighter ground. Overgrain sides to meet the dark pencilled work, taking care not to show a joint or lap. When dry, the work can be lightly overgrained in oil, accenting the shadows or mottled places in the work. As a rule, these seldom appear in the natural wood; the general effect is quite plain, the variations in color being the chief characteristic of the wood.

In oil graining the work is rubbed in with a mixture of the colors named above, the color being of a light shade and spread out rather sparingly. The rag can now be used to indicate the general direction of the grain. This is done by folding the rag loosely in the hand and describing on the work the direction of the figure. Have a little of the rubbing-in color darkened with burnt sienna and lake and slightly thickened with bolted whiting. This color is applied with the flat fresco bristle liner to produce the figure, and is immediately blended with the rubbing-in brush or with a badger blender. Combs may be used to carry out the grains on the sides of heart grains or an oil overgrainer may be used for this purpose. Some of the rift grains are extremely plain and some of the figure work quite bold; it depends on how the timber is cut. A slight overgrain in oil or water gives depth and transparency to the work. Some varieties show mottled or bird's-eye figure, and can be imitated in the manner directed for maple or cherry.

There is considerable difference in the color of different boards of this wood; the colors range from very light to very dark. The light shades are similar in color to cypress or light ash, generally of a warm tone, and the dark shades are a rich reddish brown approaching to black.

CHAPTER XXV

YELLOW PINE

Ground-color. — White lead, medium chrome yellow.

Graining Color. — Yellow ochre, raw sienna, burnt sienna, rose pink.

Tools for Oil Color. — Rubbing-in brush, sash tool, fitches, bristle fresco liner, combs, rags, etc.

Tools for Water Color. — Sponges, rubbing-in brush, sash tool, fitch tools, bristle liner, overgrainers.

This wood is a native of the southern portion of the United States. It is often the drained pitch pine or long-leaved pine tree which has been killed by taking away the sap to make turpentine.

This wood is used much more frequently in the United States than formerly, partly because of the scarcity of white pine and partly because of its boldness and variety of figure. For the latter reason it is finished in varnish, and often takes the place of white pine, which formerly was painted and grained to imitate oak in kitchens and rooms of ordinary dwellings. It is difficult to make a door of this wood which will stand extreme changes of temperature. The joints open and look badly in a comparatively short time, so that often doors of another wood are used, and are painted and grained to represent the yellow pine. No door stands atmospheric changes so well as one made from white pine, unless we except the doors which we are promised in the near future which are to be made of compressed wood-pulp without joints, the mouldings being compressed with the door. This wood has a figure very similar to that of Norway pine, and the colors used are similar; for some of the warm, soft tones a little crimson lake must be used, also a very little drop-black, but care must be taken not to produce a greenish tone if the black is used over a bright yellowish ground-color. The lake is best used as a shading color to be applied over the work in a thin wash to bring it to the desired shade. The faint mottled effect peculiar to this wood is obtained in a similar way, using the mottler in water color, or a soft rag if oil is the vehicle.

This wood may be represented in either oil or water color, but my preference is for the former. First, because it can be more quickly rubbed in and also because you do not have to wait for the weather should the temperature be below freezing.

Mix the graining color with a little raw sienna, a little yellow ochre, and a touch of burnt sienna; thin with the thinners previously described. The work is combed with a rubber comb, and the grains applied with a fitch tool or bristle liner.

PITCH PINE, OR HARD PINE

Ground-color. — White lead, chrome yellow, venetian red.

Graining Color. — Raw sienna, burnt sienna, burnt umber.

Tools. — Rubbing-in brush, sash tool, fitch tool, bristle liner, overgrainer, combs, rags, etc.

This wood is the long-leaf pine of the South Atlantic and Gulf states, and from its sap is made the spirits of turpentine.

This wood is used for floors more than any other wood in the United States, and for this purpose is generally sawed in a manner similar to quartered oak. The rings of annual growth are intersected by the saw as nearly as possible at right angles, so that the grains run very nearly parallel and wear much more evenly for floors than when cut with a heart figure.

For standing finish the wood is sawed in the usual manner, and the grains are often very bold and strong, showing strong contrasts. It is sometimes necessary to add a little venetian red to the pencilling color to make it sufficiently dense and opaque to match the dark heart grains.

Prepare a mixture of one-third raw sienna, one-third burnt sienna, and one-third burnt umber, and thin it to a very fluid consistency — in fact, a mere wash. This is for the rubbing-in color. Add to this color about one-half gill of raw linseed oil for each half pint of color which has been thinned with the regular thinners. Spread this color rather sparingly on the work. Then take some thick color, made mostly of burnt sienna and burnt umber (with a touch of venetian red if necessary), and pencil in the strong heart grains, using the flat fresco bristle liner; blend with the rubbing-in brush. It is sometimes necessary to remove nearly all the rubbing-in color with a rag before beginning to pencil in the color. In fact, excellent work can be done by pencilling the color on the dry ground work, and when dry, rub in lightly and comb or mottle the work whenever necessary.

When the work is to be combed, it will be necessary to slightly thicken the color. An overgrainer may successfully be used in oil colors. Fasten a bone comb, teeth upward, on the inside edge of the pot, the teeth of the comb being slightly above the edge of the pot. Dip the overgrainer in the color and draw it through the teeth of the comb; this will separate the bristles so that the color can be evenly applied. The piped overgrainer can also be used with success.

To produce the best effect the work, when dry, should be overgrained, using some very thin color and if necessary adding a little crimson lake and drop-black to the overgraining color.

This wood can be well represented in water color; but my preference is for oil color, as the heart grains are bold and prominent and ample time is allowed to blend the work in oil color without drawing up the color to a thin, sharp edge.

CHAPTER XXVI

CYPRESS

Ground-color. — White lead tinted with raw and burnt sienna.

Graining Color. — Raw sienna, burnt sienna, and drop-black.

Tools for Oil Color. — Rubbing-in brush, sash tool, bristle liner, overgrainer, combs, rags, etc.

Tools for Water Color. — Sponge, rubbing-in brush, sash tool, fitches, overgrainers, blender, and bristle liner.

This wood grows plentifully in the southern portions of the United States and is being used for interior finish much more frequently, owing to the scarcity of white pine and its increased cost; it is also much more in evidence than in former years, largely for the reasons stated for the increasing use of yellow pine. Some of its figures are very beautiful; but it is a very soft and porous wood and easily bruised or dented, and hence is not as desirable for interior finish as pine. We rarely find it properly finished, being composed of alternate layers of soft and hard fibres. The grain will lift unless it is properly filled in the beginning, and this filler should be a hard, non-porous substance and sufficiently transparent not to obscure the natural beauty of the wood.

The grain of cypress resembles that of hard pine, but is broader in the heart and finer grained; there is also more contrast between the light and dark portions of the growth. The ground is slightly darker and more yellow than that used for oak. The graining color is made of raw and burnt sienna and burnt umber, and is mixed in oil. When the color is rubbed in, the hearts are wiped out in the usual manner or pencilled in. A rubber comb can be used to make portions of the heart by occasionally using it in the finer portions of the wiped-out hearts. Great care should be taken that the lines made by the comb closely follow those made by hand, and that they are equally distinct. The fitch tool is often used in matching cypress; the combing is mostly fine and rather straight. Never use the steel combs over the lines made by the rubber comb. The work may be shaded with some of the graining color to which some black has been added and the whole thinned with turpentine, but the work is ordinarily finished without shading.

CHAPTER XXVII

QUARTERED SYCAMORE

Ground-color. — White lead, raw sienna, burnt sienna.

Graining Color. — Raw sienna, burnt sienna, and drop-black.

Tools for Water Color. — Sponge, sash tool, rubbing-in brush, stippler, blender, bristle liner, camel's-hair pencil, crayons.

Tools for Oil Color. — Rubbing-in brush, sash tool, fitch tools, flat fresco bristle liner, camel's-hair pencil, crayons.

This wood grows plentifully in the northern portion of the United States and is found in nearly all temperate climates. It attains great height and girth, but with age it invariably decays in the centre. It is found plentifully in Ohio, Indiana, and Illinois. Senator Voorhes of Indiana was called by his colleagues at Washington "the tall sycamore of the Wabash."

If the tree is sawed in the ordinary manner, it presents a very subdued grain and scarcely any figure is shown; but when cut from the centre of the log toward the bark, or "quarter sawed," as the lumbermen call it, the figure produced by the medullary rays is exceedingly beautiful and is very difficult to imitate, as it is full of minute grains which cannot be ignored if a faithful imitation is to be produced.

My preference is for water color if the work is to be well done. Use one-fourth stale beer to three-fourths water, or the same proportions of vinegar and water. First dampen in the work with the sponge or rubbing-in brush, use dry whiting if it crawls, and take sparingly, on a sash tool, a little raw sienna and a touch of burnt sienna; add a little drop-black, or keep the black in a separate fitch tool and blend in with the other colors. A background of shaded veins must be made similar to those in mahogany, but much less pronounced. Draw the blender lightly through these veins and blend lightly across the grain. This should produce a background effect for the fine veins which are formed by the medullary rays in the natural wood. The stippler or the badger blender may be lightly used across the grain to suggest the fine quartered veins. In some pieces of the work the longitudinal streaks or veins may be omitted and a cross-stippled background produced, but it must be done with very thin color and present a faint suggestion of the quartered veins.

When this has been allowed to dry, the dark veins, which make the beauty of the grains, are put in with a brown or a reddish brown crayon pencil. The work can then be varnished before being overgrained, or it may be overgrained in oil color directly over the water color. Care must be taken to have the brushes clean and the color very thin, merely a wash of one-third linseed oil to two-thirds turpentine, adding a sufficient quantity of liquid dryer. The brush should not be too much worn or it will be coarse and rub up some of the fine crayon markings. A camel's-hair pencil may be used in water color or in oil color to put in the fine quartered grains, but the crayon process is by far the most rapid.

In doing the work wholly in oil the groundwork should be nearly flat. The graining color is prepared very thin and is composed of a very little raw sienna, burnt sienna, and drop-black. When a panel is rubbed in, take a sash tool charged with some darker color and block out the background of longitudinal veins. They must not be made too dark. Two or three shades darker than the rubbing-in color is sufficiently strong. Then take a soft rag and remove portions of the color between the dark veins; blend the panel lengthwise with the rubbing-in brush and then lightly crosswise and always in one direction. The crayon pencil can then be used, and if the ground-color has been properly prepared and applied, the dark veins can be represented with considerable accuracy. If the ground-color is too glossy, the crayon color will not adhere to the work and the color must be applied with a very small bristle fitch tool or with a camel's-hair pencil. The work should, in all cases, be overgrained if the best results are to be obtained. The overgraining color being only a thin wash to give depth to the work, dark veins can be produced in the overgraining, or those already put in can be accented if necessary.

Especial care should be taken in imitating this wood to have the bolder figures appear in the panels, and the grains in the stiles, especially in the long stiles or rails, should be more subdued. It is well to use crayons of different shades, putting the darker figures on the panels, and, in case the panel is very large, using two or more crayons on the same panel.

By close observation of the real wood we will find that often in the same panel there appear streaks or veins of a lighter or darker shade from those on either side. We can represent this peculiarity by using different colored crayons, or, in case we are using the camel's-hair pencil, by darkening the color or by making it lighter.

Quartered sycamore is seldom imitated and a good imitation might cost nearly as much as the real wood, as a man could easily spend a whole day on one side of a door if the work was to match some fine specimen of the wood.

CHAPTER XXVIII

CHERRY

Ground-color. — White lead, yellow ochre, venetian red.

Graining Color. — Raw sienna, burnt sienna, burnt umber, drop-black.

Tools for Oil. — Rubbing-in brush, sash tool, fitches, combs, rags, bristle liner, overgrainer.

Tools for Water Color. — Crayon, sponge, rubbing-in brush, sash tools, fitch tools, blender, bristle liner, overgrainer.

This wood grows throughout the northern portion of the American continent, and when finished natural is often as light in color as ash and its markings are very subdued and quiet in character. It is seldom finished in its natural color, as the popular idea of cherry color is one much nearer to the color of the fruit than of the color of the wood, so that it is more frequently stained to a mahogany color, and being a dense hard wood it can be most successfully stained to any depth of color.

Furniture made from this wood and stained is frequently sold for mahogany, and it often requires a close inspection to detect the deception.

Cherry may be well imitated in either water color or oil. The base of color for a natural cherry is raw sienna deepened with a little burnt sienna and a touch of drop-black. If a very light shade is desired, burnt umber will darken the color sufficiently. If the color is to be a deep reddish shade, the base of the graining color will be burnt sienna deepened with burnt umber, adding a little black if necessary. The color should be applied rather sparingly, if oil color is used, as most of the figure is pencilled on the work with the bristle liner and if there is a surplus of color on the work, it will not blend successfully.

For oil color, rub in as usual; then take a little of the graining color and darken it slightly by adding dry burnt umber. Mix the color in a separate tin and use it to apply the figured grains to the work. The flat fresco bristle liner is an excellent tool for this purpose, although some grainers prefer a camel's-hair pencil.

The mottled effect of cherry may be obtained in oil color by using a fitch tool with a little of the rubbing-in color or a little of the darkened pencilling color, and with it apply the mottled dark markings in the general direction of those seen in the wood. Blend lightly crosswise with the rubbing-in brush in the general direction of the mottlings; then blend lightly one way, lengthwise of the grain. If a pencilled growth is to be applied over the mottling, care must be taken to blend it in the same direction, lengthwise, as the mottling, otherwise the effect of the mottling is largely destroyed.

The color should be allowed to set slightly before either mottling or pencilling is attempted, as if done too soon the colors will blend together and the sharp and clean effect of the grains will be lost. On the other hand the work must not be allowed to set too much, as the wet color will lift the grain too much and the effect will be too pronounced and unnatural.

The piped overgrainer, or the short-haired overgrainer, may be used in oil very successfully in imitating this wood. Mottle the background as directed, and when slightly set, use the overgrainer in oil color and blend quickly with the rubbing-in brush.

GRAINING, ANCIENT AND MODERN 83

Combs can be used to represent the light, plain portions of the wood, the medium and fine steel combs being best suited for this purpose. The straw matting may also be used and will be found useful in carrying out the fine lines on either side of a panel where the pencil has been used to put in the heart grains.

Some of the finer heart grains may be wiped out with a rag, but the wood can best be represented by pencilling in the color. In fact, the heart grains of nearly all the hard woods can be most successfully imitated by pencilling rather than by wiping out the color. There is, in cherry, a fine secondary grain which must not be ignored.

The mottled effect of cherry wood can be very closely represented by using a mottler in water color on the groundwork and overgraining in oil color. In fact, the whole of the work can be done in water color, and an oil glaze of the proper shade gives it the necessary depth to make a very natural appearance.

A close examination of this wood will reveal the fact that its heart grains are very finely outlined, and a very small fresco liner and rather thin color should be used or the grains will be made too prominent. The crayon pencil encased in wood, or the soft crayon of home manufacture, may be used with success for all heart grains in both oil and water colors.

In graining this wood wholly in water colors the work is first sponged over with stale beer diluted with one-half clean water; then apply the graining color very thin, and while wet lay out the mottled effect, using the fitch tool with color slightly darkened, or gathering the color with the mottler and blending lightly with the badger blender. When the mottling has dried, the grains can be put in with the bristle fresco liner or with the piped or short-haired overgrainer. Care must be taken to blend the work quickly and draw the color to a sharp edge on the outside of the heart grains. By using two-thirds stale beer in the mottling color, and little or none in the pencilling color, the underneath color is not so readily softened by the wet pencil, and if carefully blended will not lift the underneath color to any great extent.

Some grainers use an oil pencilling color over the water-color mottling. Thin the pencilling color with raw linseed oil and benzine to which a little liquid dryer has been added. This color does not blend so well as the water color, but has the merit of lessening the danger of rubbing up the dry mottled color.

To obtain the best results, the work, whether done in oil or in water color, should be overgrained. For a very bright shade a little crimson lake with a touch of drop-black may be added to the regular oil color and the whole thinned forty to fifty per cent with turpentine. Give the work a very thin coat of this color, taking pains to spread it evenly and quickly, being careful that it does not rub up the underneath color, if the latter has been done in oil.

In dismissing this wood from further consideration, a word of advice to young beginners will not be out of place. Try to make a difference in both the character and the relative thickness of the grains of cherry from that of any other wood except birch or maple. The most common fault of imitations of cherry is that the heart grains are made too prominent as well as too thick. Look at the wood and see how fine some of the heart grains appear; nothing less wide than a fine camel's-hair pencil or a fine crayon could be used to represent such grains.

A faint stippled or porous effect is often observed in some varieties of cherry. There are also some portions of the tree which show, when cut parallel to the medullary rays, a very pronounced quarter-sawed effect similar in character to quartered sycamore. But the veins in cherry, as a rule, are light on a darker ground, while those of sycamore are, as a rule, just the opposite. A similar effect is produced by using the stippler to make faint checked markings across the grain, stippling lightly across the panel, having very little color on the work. This should only be done on the work at rare intervals. The faint porous effect obtained by stippling in the usual manner is much more frequently seen. The stippling color should be very thin or the effect produced is liable to suggest mahogany rather than cherry.

CHAPTER XXIX

CURLY BIRCH

Ground-color. — White lead, raw sienna, venetian red.

Graining Color. — Raw sienna, burnt sienna, burnt umber, drop-black.

Tools for Oil Color. — Rubbing-in brush, sash tool, fitch tools, flat fresco bristle liner, combs, rags, piped and short bristle overgrainer, crayons.

Tools for Water Color. — Sponge, rubbing-in brush, sash tool, fitch tools, flat fresco bristle liner, blender, overgrainer, crayon.

This wood grows in northern North America and in nearly all temperate climates. It is very similar to cherry in general character and often is stained to represent stained cherry. Being of about the same density, it makes a very successful imitation. The mottlings of curly birch are invariably more pronounced than those of cherry, and if the wood is stained, the mottled effect is very positive, on account of the stain penetrating the ends of the pores of the mottled wood, which readily absorb it.

If the wood is to be represented in its natural color, three-fourths raw sienna and one-fourth burnt umber may be used, thinned to a transparent stain; but if a stained effect is wanted, burnt sienna and burnt umber must be used;

also a little drop-black, the shade of color depending on the color of the wood to be matched.

For oil graining the work is first rubbed in and allowed to set slightly, then the mottled effect can be produced by using the fitch tool in color slightly darkened with a little burnt umber. This color is lightly blended with the rubbing-in brush and allowed to set still longer. Then pencil in the growth with the bristle liner, blending one way. Carry out the side lines of the heart grains with combs or with the overgrainer.

The piped overgrainer or the short-haired overgrainer may be used in oil over the mottling with excellent effect.

There is, in some varieties of curly birch, a warm pinkish tinge which cannot successfully be matched with burnt sienna. In such cases a little rose pink or crimson lake may be used to advantage. A little drop-black should be mixed with the lake and the color thinned to a very fluid consistency; it is best used as an overgraining or glazing color, although a little of the lake may be added to the graining color. The pinkish portions of the wood are found more rarely than the reddish brown to gray toned portions, hence it would not be wise to stain all the color to a pinkish tone; but where such pieces are desired, use a little of the lake or pink to brighten the color already applied to the work.

Before the heart grains are put in with the pencil, brush, or bristle liner it is well to outline with a soft rag the general direction in which the heart grains are to run. Fold the rag several times and with sweeping strokes wipe off portions of the color, so that when the pencilled work is put in the color will not run. If mottling is to be done on such portions, it should be done before the pencilling is applied.

To represent this wood in water color, the colors to be rubbed in should be kept in separate vessels and a little of each color taken up in the large sash tool. This can be spread out with the rubbing-in brush and a mottler or fitch tool used to represent the mottled curly markings. Care must be taken to have the mottling irregular and not running too parallel. The mottlings of curly birch, while very pronounced, are also very irregular, and the color must be broken up into patches of strong mottles intersected with bright portions of the wood. When the mottling is dry, the heart grains can be put in with the bristle liner or with the camel's-hair pencil and blended to bring the dark edges of the figure on the outside of the grain.

The piped overgrainer or the short-haired overgrainer can be used to overgrain the mottled work or to put the grain in the plainer portions of the work. A thin glaze of overgraining color is necessary to finish the work whether it be done in oil or in water color.

CHAPTER XXX

BLACK WALNUT

Ground-color. — Yellow ochre, white lead, venetian red, burnt umber.

Graining Color. — Burnt umber darkened with Vandyke brown.

Tools for Water Color. — Sponge, rubbing-in brush, stippler, mottler, blender, overgrainer, pencil fitch, sash tool, bristle liner, crayons.

Tools for Oil Color. — Rubbing-in brush, sash tool, fitch tool, flat fresco bristle liner, combs, rags, crayons.

This wood is a native of the middle portion of the United States and of the southern portion of Canada west. It is now seldom used for finish in the East, owing to its scarcity and consequent high price.

When the proper ground-color has been applied and

allowed ample time to dry, the work should be lightly sandpapered and stippled in water color. Use Vandyke brown ground in water and burnt umber in equal parts. For a lighter shade of walnut the Vandyke brown may be omitted. The colors ought to be thinned with not more than one-third to one-half stale beer to one-half to two-thirds clean water. This is amply strong to bind the color to the work. The figure can then be put in by using fitch tools and overgrainers directly on the work as stippled, which is the method preferred by many artistic workmen. Or, when dry, it may be rubbed in in oil color, using straight burnt umber for the graining color thinned with one-third raw linseed oil to two-thirds spirits of turpentine, adding about one gill of good liquid dryer to the half gallon of thinners. The grains can then be put in with fitch tools and overgrainers. Combs can be used and the work finished, or it may be wiped out with a soft cotton rag. Some old grainers prefer graining the work in oil, and stippling in water color when the oil color is dry.

In some parts of the country the colors used for black walnut are burnt sienna and lamp black in varying proportions, but with such colors it is easy to produce striking contrasts rather than a good imitation of the wood. There is no doubt that a skilful workman can do a good job with these colors, but the average shade of black walnut is much nearer the tone of burnt umber than that of burnt sienna or black or either of them in combination.

The average grains to represent heart growths are much too strong, nor is there the contrast between the edges of the figure and the general tone of the wood that some grainers seem so anxious to produce. The quiet and subdued growths are far the most numerous. The bolder figure challenges attention, and is more readily discovered to be an imitation.

The custom of plentifully besprinkling imitations of American walnut with knots is most absurd. It is true that an occasional knot may be found in the wood, but some grainers seem to delight in putting more knots into one walnut door than could be found in a cart load of walnut timber. A knot may be beautifully imitated, but it is generally considered an imperfection in the natural wood, and bears the same relation to an otherwise fine panel that a wart or mole would on the face of an otherwise beautiful woman. There are many lights, shadows, and curly places in the wood that are not imperfections, but which add to its beauty, and these can be represented with good taste, and the imitation of knots can be left to the amateur.

Sharp color contrasts are to be avoided if the repose and general effect of the work is to be considered. It is always permissible to make a slight difference in the color of the stiles and rails of a door, but not a violent contrast. We seldom see such effects in a hard-wood door, and no intelligent joiner would, of his own choice, put into one door such various colored pieces of wood as we sometimes see imitated by clever grainers.

Then again, some men will always want to have the mouldings a lighter or a darker shade than the rest of the door, so that in conjunction with the other shades the effect produced is far from reposeful, and although finely executed, may lack the primary suggestion of being natural wood, which should be the first consideration in all imitations of wood.

The foregoing remarks as to color effects may apply with equal force to all woods, but particularly to walnut or other dark woods.

I have heard of two carpenters who spent a week in sorting over oak timber, to which a coat of oil had been applied, in order to select the wood as nearly as possible of the same shade so that one large room should be finished in the most artistic manner. This is quite opposite in effect from that which some grainers strive to obtain when they make the color contrasts we frequently see in one room. It is wise to refrain from extremes in graining, as in other things, and a temperate course is, in the main, apt to give the best satisfaction. The work should not look insipid nor lack character; still, it need not offend by presenting too violent contrasts. Aim at quiet and reposeful effects rather than at pronounced and glaring work. Keep all joints and divisions clean and the effect will be repaid by a more woody appearance.

CRAYONS FOR WALNUT GRAINING

A very good imitation of black walnut may be produced by the use of crayons in either water or oil color. If used in water color, the stippling is first done in the usual manner on the groundwork, and when dry the crayon is used on the dry stippling to put in the heart grains. The outline should not be too bold and the fine lines at the open ends of the growths can be carefully put in. A soft piece of rag or a stiff dust brush should be used to slightly soften the harshness of the lines, as left by the dry crayon, otherwise the work is apt to appear too bold. An overgrainer charged with water color, the bristles having been divided with a bone comb, may be used to carry out the lines, on either side of the hearts, to the edge of the panel. Care must be taken to have the color in the overgrainer as nearly as possible of the same shade as that of the crayon. It may be kept a little lighter rather than darker. When dry, glaze over with a thin wash of oil thinner, using a small portion of burnt umber in the color. If the crayon is used in oil, the work must first be grounded in rather flat; when thoroughly dry, rub in with very thin color mixed about one-half oil and one-half turpentine, adding a very little dryer. Then outline your hearts with the crayon; blend and fill in the sides with overgrainer in oil, or rub in the sides of panels with a little darker color and use combs.

GRAINING, ANCIENT AND MODERN

When dry, the work ought to be shaded or glazed in oil, using one-third oil to two-thirds turpentine and very little dryer.

CURLY WALNUT

Ground-color. — Yellow ochre, white lead, venetian red.

Graining Color. — Burnt umber darkened with Vandyke brown.

Tools for Water Color. — Sponge, rubbing-in brush, stippler, mottler, blender, overgrainer, pencil fitch, sash tool, bristle liner.

Tools for Oil Color. — Rubbing-in brush, sash tool, fitch tool, flat fresco bristle liner, combs, rags.

This wood, in America, is a fine variety of black walnut, and grows in the same territory. It is generally obtained from an old tree, but the effect may be produced by a method of cutting veneers from a log on the outside, always following the circumference of the log. The American variety is less pronounced in its lights and shades than the European, and the colors, as a rule, are much lighter. The same colors used for black walnut will answer for this wood, and the same general method of treatment will suffice. The strong shadows across the grain may be put in wholly in the oil color, or they may be done in the water color and accented in oil after being overgrained. In all cases the work should first be stippled in water color to produce the effect of the pores of the wood.

FRENCH WALNUT BURL

Ground-color. — Yellow ochre, white lead, venetian red.

Graining Color. — Burnt umber darkened with Vandyke brown.

Tools for Water Color. — Sponge, rubbing-in brush, stippler, mottler, blender, overgrainer, pencil fitch, sash tool, bristle liner.

Tools for Oil Color. — Rubbing-in brush, sash tool, fitch tool, flat fresco bristle liner, combs, rags.

This wood, as its name implies, is grown in France, and is secured from excrescences or burls which form on the side of the tree. These are sawed or cut, and the grain produced is very fine.

This wood was formerly very much used on the better class of furniture, for panels, and it was sometimes used on doors. Being, as a rule, but a thin veneer, it is unsuitable for use in exposed situations or in any climate of extreme temperatures. The burl is a wart or excrescence which is cut from the side of the tree and sawn or sliced into thin veneer.

French walnut may be represented in either water color or oil. Most grainers prefer the former method, as the work can be executed more expeditiously and overgrained at once. The tools are the same as those used for black walnut, as are also the colors — burnt umber and Vandyke brown. For the very light portions a little burnt sienna may be added to the color. If the work is to be done in oil, rub in the color rather dry, and with the sash tool dipped in some dark color cover the portions of the work which you desire to appear dark; then with a piece of soft rag remove the color where the light places are to appear, and work up the dark places with the rag until the desired effect is obtained. Blend lightly with the dry brush, and add lines and curves with the fitch tool. Then stipple the light places with the flat brush. When the oil color is dry, the work may be shaded or overgrained in either oil or water color.

If the work is to be done in water color, use a sponge for blocking out the lights and add dark color with a fitch tool; mottle and blend lightly, and with the fitch tool and overgrainer put in the grains over the mottled work. When dry, it may be lightly varnished and overgrained in water color.

A careful study of the grains of this wood will show that the darker lines or veins are sharp on one edge and softened on the other. Or, in many cases, a vein seems to come from below the surface, and appears only as a sharp edge on the work. Do not have the figures too bold. Vary the work and reverse it in opposite panels, as that is the method most frequently adopted by the joiners in constructing the natural wood.

Years ago this wood was frequently represented on the panels of outside doors, but of late years in this vicinity very little French walnut is imitated. Dark oak is the reigning favorite for outside work.

ITALIAN WALNUT

Colors and tools the same as for black walnut.

This beautiful wood is seldom seen in this vicinity and is very seldom imitated. Its grains have many features similar to those of French burl walnut.

It can be represented in water colors. It will be necessary to use a little drop-black in the graining color, mostly in the pencilling color, as these grains appear quite dark on the lighter background.

Rub in the work and with a sponge block out the general direction of the grains. Follow these with the fitch tool and blend lightly, always using the badger blender for softening the harsh look of the work. The mottler should be frequently used. When dry, overgrain lightly and blend.

CIRCASSIAN WALNUT

Colors and tools the same as for black walnut.

This wood is found in the southern portions of Europe, and is usually a thin veneer applied to a background of inferior wood.

It is best represented in water colors, and when dry may be lightly overgrained in oil. A little drop-black may be necessary in making the darkest lines or veins, but, as a rule, the lines while very pronounced are very thin and should be drawn to an edge with the blender. The tone of the color of the wood suggests the use of more black in our color if we would faithfully represent its natural color.

CHAPTER XXXI

MAHOGANY

Ground-color. — Yellow ochre, orange chrome, venetian red, red lead, and white lead.

Graining Color. — Burnt sienna, rose pink, Vandyke brown, crimson lake.

Tools for Water Color. — Sponge, rubbing-in brush, stippler, mottler, cut tools, sash tool, fitch tools, bristle liner, overgrainer, blender.

Tools for Oil Color. — Rubbing-in brush, fitch tools, sash tool, bristle liner, overgrainer.

This wood is a native of America, and some of its most beautiful timber is sawn from logs cut in Honduras and Mexico. It also grows in Cuba and in central and northern South America. There are also several varieties which are found in Asia and in Africa.

The old feather-grained mahogany of Honduras is probably the most beautiful of all the varieties and is exceedingly difficult to imitate. I have been told that the wood at the butt of the old Honduras mahogany tree is so dense and so difficult to cut that the natives build a platform around the tree some distance from the ground and cut the tree from that point owing to the wood being softer and more easily chopped away.

Of the many varieties of mahogany that are used for furniture we find but few that are used for interior finish.

GRAINING, ANCIENT AND MODERN 95

These are generally of the lighter shades of the wood and frequently they lack any bold figure. Such mahogany is most frequently stained in color, to represent the more valuable timber from which the best furniture is made.

This variety is not so difficult to imitate as the mottled or feathered variety. A good background and proper stippling are the essential elements of success. Prepare the stippling color with Vandyke brown ground in water, and thin with one part stale beer to three parts water. Stipple in all portions of the work intended to be finished plain; where the feather is to be represented, the main part of the work must be done in water color. After the stippling is dry, the oil color may be applied; this is made by mixing one part burnt sienna, one part rose pink, and one part Vandyke brown thinned with the regular oil thinners. Apply this color evenly. Then take some Vandyke brown thinned with turpentine and liquid dryer (as this color is one of the slowest to dry), and with a fitch tool put in the darker veins. Blend with the flat brush lengthwise, then lightly crosswise, drawing the color to a sharp, dark edge on the sides of the darker veins. If necessary, the flat brush may be used as a stippler in the oil color and the work lightly stippled across the grain or lengthwise as may be preferred.

If the mottled effect is desired, it may be produced by the use of a small fitch tool in the thin color, making the mottlings across and between the darker veins, or the stippling color may be bound down with a thin coat of varnish and the mottling done in water color on the dry varnish.

The short bristle overgrainer may be used to overgrain the work in oil, and when so used, the work must be blended lightly across the grain and always in one direction.

The feather veins of mahogany can be represented wholly in water color; when the panel is sponged in, the work may be rubbed in with the rubbing-in brush. Then darken the centre of the panel where the feather is to be represented, using the clear Vandyke brown for this purpose. Use a sponge to remove portions of the color in the general direction of the finished work; then with the cut tool or mottler, work out the general direction of the feathered work and blend in the edges of the darker veins. If the color dries before the work is satisfactory, it can be carefully wet over with clean water applied with a short-haired overgrainer and the cut tool or mottler again used to bring out the bright portions of the work.

It is impossible to describe with accuracy the exact method to produce this intricate and complicated wood. Nothing but a careful and painstaking study of actual examples of the wood will give the student the proper ideas of how it ought to be represented. When the mottled and veined effect of the feather grain is dry, it should be overgrained with some thin Vandyke brown, using the piped overgrainer or the short-haired overgrainer. The latter is preferable, as the hair can be divided into irregular portions by the bone comb and the work will look less mechanical than if done wholly with an overgrainer with one width of lines. The piped overgrainer may occasionally be used in connection with the other overgrainer, but care must be taken to have the color in each brush of precisely the same shade on the same panel, otherwise the work will look patchy. The color must be quickly applied with the overgrainer, following the general direction of the grains previously done, and the badger blender must be used at once to draw the overgrained work to dark edges similar to the figure seen in the natural wood. Have the light and dark veins well laid out in the primary stages of the work and take pains not to get the work too dark, as it can easily be brought to any depth of color by overgraining, or shading; but if too dark at the beginning, it cannot well be lightened without repainting.

In doing fine work it is best to apply a thin coat of

varnish to the work after the first graining, and when dry, apply the overgraining color, or, if necessary, first go over the graining again and accent any of the primary work. When the water-color work is dry, it may be lightly glazed over in oil color, using a little crimson lake in the color. This will give both depth and brightness to the work. Use but little oil in the color, mostly turpentine and a little liquid dryer if necessary. The darker shades can be again worked over and details of figure added. It will take many attempts and an abundance of patient work before even a tolerable success can be gained in the imitation of feather-grained mahogany. Secure, if possible, a good specimen of the real wood and have several panels grounded in color similar to the lightest shade in the real wood, then practise to obtain the effect of the light and shade of the wood. Try some panels with a light stippled background done with thin Vandyke brown in water color. When dry, give them a thin coat of varnish, and when that is thoroughly dry and hard, grain in water color as previously directed. Notice whether the effect of the stippling is too pronounced by comparison with the work done on the panels without stippling, and compare both panels with the wood. You will then be able to determine whether your stippling on the groundwork is an improvement over that done without stippling. The stippling can be added to such portions of the work as appear to need it by applying the stippling color with a round blender or with a round or oval sash tool, not first rubbing in the work, but having some color in the brush and lightly touching the sides of the brush against the places to be stippled.

CHAPTER XXXII

TEAK

Ground-color. — Yellow ochre, venetian red, chrome yellow, and a little white lead.

Graining Color. — Burnt sienna, Vandyke brown, and a little rose pink.

Tools for Oil Color. — Rubbing-in brush, sash tool, fitch tools, bristle liner, piped and short-bristle overgrainers.

Tools for Water Color. — Sponge, rubbing-in brush, stippler, sash tool, fitches, bristle liner, blender, overgrainers.

This wood is a native of India and has a grain somewhat similar to that of mahogany. It resembles the latter wood in many particulars. It is a very hard, dense wood, holding its color well and taking a high polish. It is used principally on steamships for covering of sides or tops of deck-houses, cabin doors, etc. It stands exposure to the weather much better than mahogany.

It may be represented wholly in water color by first stippling on the groundwork a thin wash of Vandyke brown ground in water and thinned with one-third stale beer to two-thirds water. The graining color is composed of Vandyke brown and burnt sienna in equal parts. The grains of the wood may be applied to this background with the fitch tool or bristle liner, and the lines on the sides of the heart grains may be carried out by using the overgrainer charged with the same color as the liner.

The work must be blended at once, using the badger blender, taking care not to lift the color too much from the stippled background. When dry, it may be overgrained in oil color with a little of the rose pink added to the graining color and the whole thinned with turpentine.

For oil color graining it is best to first stipple the work

in water color to produce the porous effect, and when dry, rub in the graining color in oil. The color should be composed of about equal quantities of burnt sienna and Vandyke brown, adding a little rose pink if necessary.

When this has been properly applied, the heart grains may be put in, using the bristle liner for this purpose and blending the color with the rubbing-in brush. Many of the grains of the wood are extremely plain, showing no sign of any other grain than the strong stippled or porous effect.

Teak is seldom imitated in Massachusetts, and when done, is generally on some transatlantic steamer.

The work may be done wholly in oil colors, but the stippled effect is more successfully produced by using water color for the under coat.

CHAPTER XXXIII

ROSEWOOD

Ground-color. — Orange chrome, red lead, white lead.

Graining Color. — Vandyke brown, rose pink, drop-black.

Tools. — Sponge, rubbing-in brush, sash tool, fitches, mottler, overgrainers, camel's-hair pencil, black crayon.

This wood grows in tropical climates. The best specimens come from Africa. It is a very dense and close-grained wood. Its color varies from a light orange to a jet black. There are probably more variations of color in this wood than in any other that is used for finish or furniture. It is very seldom used as a finish for rooms, and in thirty-three years' experience I have seen but one rosewood door and frame used on any building on which I have worked.

The grains of this wood when finished natural are very beautiful. The figures run in streaks or veins and seem to interlock in most eccentric fashion. The outer edges of these veins are frequently black and stand out prominently on a lighter background. In some of the veins of figured work the background assumes a pinkish tinge which is well imitated by the use of rose pink.

The color is first applied sparingly with the sponge or flat brush, using diluted Vandyke brown for the color. One-third stale beer or vinegar to two-thirds water will suffice to bind the color to the painted surface. Remove portions of the color with the sponge and put in veins of drop-black with a fitch tool. The blender is then used to soften the outlines and it may be drawn through some of the veins inside the outer edges. If it is desired to have some of the veins of a pink shade, a sash tool charged with rose pink is used to apply the color. The background of some of the veins may be faintly stippled, or the stippling or checking may be done in the overgraining color. The black crayon pencil does excellent work for this purpose.

When the outlined veins are dry, the overgrainer (either piped or short haired) is charged with diluted drop-black and the color is applied in the direction of the fine grains seen in the wood; the overgraining is immediately blended with the badger blender. If carefully done, this will draw the color to sharp edges and produce an effect very similar to the grain of the wood. Without a bright background, it is useless to attempt to grain rosewood, as the color effect depends so largely on the transparent brilliancy of the ground-color and on this brilliancy depends in no small degree the success of the work.

The camel's-hair pencil is used to put in some of the border heart grains, also to sharpen the edges of the dark veins. The work must be immediately blended.

I have read instructions for graining this wood in which

the ground-color is stated to be black, the graining color is also black, and it gives one a chance to wonder how black graining color would appear on a black ground.

CHAPTER XXXIV

OVERGRAINING

IN these days of hurry and rush, opportunity is seldom given to finish the work, and even expert grainers get into the habit of considering their work finished without being overgrained; yet it is a fact that scarcely any wood can be so well done at one treatment as not to be vastly improved by being overgrained. The light and shade, however effectively disposed at the first treatment, can be made much more effective by judiciously overgraining. A common fault, even among expert grainers, is that they try to do too much at once.

It is not the intention of the writer to disparage in any way the work of skilled men nor the processes which, by careful experiment and years of practical application, they have evolved and adopted; but it is none the less true that their work would often look better if the effects striven for in oil color were applied in the overgraining rather than in the first treatment and the time consumed would only be slightly greater.

Care should be taken in oil overgraining to have as little linseed oil as possible in the color, as an excess of oil often acts disastrously on the varnish, causing it to crack. All that can be done in oil overgraining can be done equally well in water color, but strong beer or vinegar should never be used for this purpose. It requires but very little binding material to hold the water color to the oil graining. One part beer or vinegar to four parts water will make the color adhere sufficiently to allow it to be varnished without rubbing up. If the work is done wholly in water color, one part beer or vinegar to two parts water will bind the color to the groundwork so that it can be varnished with safety.

CHAPTER XXXV

CEILINGS

CEILINGS, whether of plaster, wood, or metal, may be grained to represent wood. When a cornice is at the top of the wall, it should be included with the ceiling, or, if treated alone, it can be made to correspond with the woodwork of the room.

No more effective method of decorative treatment can be suggested for the ceiling of a modern dining room. So much oak is now used for furniture, picture frames, mouldings, etc., that a ceiling harmonizing in color effect and general character with the furniture gives a reposeful and harmonious effect which is preferable to some of the costly but rather bizarre effects sometimes produced by alleged decorations.

A ceiling properly grained will last for many years and can be easily cleaned and renovated at a slight expense.

Simple effects are often best suited to a ceiling. If flat, lay out the work in forms best suited to the surroundings. A simple plan is to divide the ceiling into four parts from the centre and parallel to the walls, then represent boards running at angles diagonally across the four squares, the boards meeting on the centre lines.

Panels may be laid out and mouldings and carved work represented in light and shade, but it is unwise to do this unless the room readily lends itself to such treatment. Much of the effectiveness of the finished work will depend on its quiet and modest appearance. Unless for special reasons it is unwise to represent a variety of woods on a ceiling; better err on the side of modesty and represent but one wood, or two at most, than to challenge attention and close inspection by imitating too many varieties of wood. When thoroughly dry, it should receive a thin coat of varnish. Use good coach varnish and dilute it with turpentine, adding a little raw linseed oil. This will allow more freedom in spreading the varnish and the finished work will not be as lustrous as if the varnish were used clear. Another reason for the use of a small quantity of linseed oil is that the varnish is applied in a thinner film, and in process of time, should cracking ensue, the cracks will be less conspicuous and finer than if a thicker varnish were used.

Dead or flat varnish is not recommended, for these reasons : it is exceedingly difficult to avoid laps in a large surface where flat varnish is used; if they fail to appear immediately after the work is done, they are likely to appear in bold prominence perhaps a year afterward. The transparency of the work is often seriously impaired by flat varnish, and if compounded on a wax basis, it is extremely difficult to clean the smoke and dust from the work without injury to the varnish, while work done with hard varnish can readily be cleaned with a diluted solution of washing soda without injury to the work. In case flat varnish is used, it is better to first apply a coat of bright varnish, and when thoroughly dry, apply the flat varnish.

CHAPTER XXXVI

FLOORS

OF late years the advent of so many quartered oak floors has made quite a demand for grained floors, and more are being done each year. If the old floor is at all smooth, and the boards sound, a very fair job can be done; but if the boards are rough, it is best to plane them before beginning to paint. If the cracks at joints are very pronounced, the floor ought to be relaid and the joints made tight. If this is impracticable, fill all cracks and openings with a mixture of rye meal and fine sawdust mixed to a paste with weak glue-size. There are several patented crack fillers for floors and they will do the work equally well.

Oil putty is not the best thing in the world for wide-open joints in a floor, as the edges of the boards absorb the oil out of the putty and the dry putty is likely to get loose.

Having properly prepared the wood for painting, the floor should receive a first coat of color with not more than one-half linseed oil and one-half turpentine with plenty of dryer. When this is thoroughly dry and hard, a second coat can be given thinned with not more than one-fourth linseed oil and three-fourths turpentine with sufficient dryer. For an ordinary floor one-half pound litharge added to the color will harden it more than liquid dryers and leave it less sticky. The floor will seldom require a third coat of ground-color; but if it does, the color should be thin and laid on smoothly. A light rub of sandpaper between the coats is essential to a smooth finish.

The floor may then be marked in narrow strips similar to the modern hard-wood floor, marking it off with a lead pencil and straight edge. Or it may be done by dividing the widest boards into two sections and leaving the narrow boards in whatever width they are laid. In the latter case mark the edges of all the boards with lead pencil. If the floor is well prepared and the joints close, it is undoubtedly better to mark it off in three or three and one-half inch strips and grain them independent of the old joints.

In most cases it is unnecessary to grain the whole floor. A border of a yard wide all around the room is sufficient, as the centre of the floor is invariably covered with a rug or carpet. This part can be stained or painted plain. Always leave a margin of ample width, allowing the carpet or rug to overlap.

On sanitary grounds alone it is worth while to paint the floors in the interior of old dwellings, and if the comparison is made between any plain painted floor and one properly prepared and grained, there can be no doubt as to the latter being by far the most beautiful, and its durability is equal to any other form of painted floor. Another thing in its favor is that when it begins to show signs of wear and tear, the worst places can be repaired and made to match the old work, a thing very difficult to do in plain paint.

The first cost of a grained floor is undoubtedly greater, but it is the cheapest in the end.

Having decided on the shade of the graining color, mix a quantity sufficient to do the job on hand, as directed elsewhere, and after the floor is all marked out begin by rubbing in the color on a strip of four or five boards on one side of the room, choosing the side on which the boards run parallel to the baseboard. Be sure the brush is well worked into the color, and rub in the boards of the same shade of color from end to end. In graining them some discretion must be used. Do not have them all of the same shade, yet make no violent contrasts. Use the combs to take off more color from some boards than from others, thus altering the depth of color. Treat the work as an intelligent carpenter would lay the floor if he had plenty of material, keeping it of the same general color and making no sharp contrasts.

In a room of ordinary size up to sixteen feet long it would be wise to make no end joints unless they appear very prominently in the old floor boards; even then it would be better to ignore them if possible. Work two-thirds or more of the distance across the floor, and then begin on the side opposite to where you first started, finishing the floor about one-third its width from the wall, allowing yourself a chance to make your exit without stepping on any of the work previously done.

Use various sizes of rubber combs and vary the combing for the background, being careful not to overlap the combing and treating each board distinct and separate from its adjoining neighbors, as would be the case in a natural-wood floor. If time allows, and expense is no object, the boards might be grained alternately and allowed to dry; then do the intervening boards, which will allow a greater opportunity for slightly altering the color and varying the style of the work without interfering with the edges of the boards already finished.

MANILA PAPER FOR COVERING A POOR FLOOR

If a floor is in rather bad condition, it may be covered with stout manila paper well pasted down after the cracks and joints of the floor have been thoroughly filled and have received one coat of flat paint. The work can then be painted as usual and laid out in three and one-half inch strips. Be sure that the paper is thoroughly pressed down on the floor so that it will adhere to every portion with which it comes in contact. Thin muslin cloth may be used for this purpose, but it is not so smooth.

The object of priming the floor with flat color is to afford an even surface for the paper, and to prevent the too-rapid absorption of the paste where the suction would be greatest on the bare wood or on cracks that have been filled with a porous material.

A fair imitation of marquetry may be made by marking out the pattern with a stencil, and filling in the pattern with the proper ground-colors; or after marking out the pattern on a light ground-color proceed to grain all the light wood in water colors. Cover all the light graining with thin asphaltum varnish or thin Damar varnish. Then grain all the dark woods in water colors on the light ground, and when dry, wipe over the work with a soft rag saturated with spirits of turpentine. This will cut off the asphaltum or Damar varnish as well as all water color over it, and if used carefully, will leave the water-color graining, both light and dark, comparatively uninjured. It is unnecessary to first remove the dark water color from the portions grained light. It will all readily be removed with the varnish by the turpentine, leaving the water color underneath clean and distinct. It is better not to allow too long an interval to elapse before taking off the stopping varnish, as it sometimes dries rather hard and requires considerable friction to remove it.

The manila paper makes an effective border treatment for marquetry, and the pattern to be grained can be marked out before the paper is pasted to the floor. In this case the paper should first receive one coat of paint. If it is desired, after the ground-color is applied, some of the figure of the quartered oak may be pencilled on the groundwork, using a thin solution of weak glue-size and sugar colored with a little dry umber and sienna. When dry, grain over it in oil, and when the oil graining is dry, the dark figures may be turned to light ones by using a clean sponge dampened in warm water and rubbed over the work. The dark figures can also be allowed to remain on the work if desired.

Another method is to pencil in the dark figures in asphaltum or Damar varnish, and when dry, grain over them with water color. When this is dry, use the rag wet with turpentine to remove the dark figures, leaving the markings light that were painted dark.

VARNISHING A GRAINED FLOOR

It is a matter of doubt what finish is best adapted to a grained floor. A coat of good floor varnish, applied several days after the oil graining is done, ought to wear and look well for a long time. If water-color graining, it may be varnished at once. If the under coats are oily and elastic, it is impossible for the best varnish to harden properly, hence it is important to have the foundation coats properly mixed and applied.

Thin shellac is sometimes used for a coating over the oil graining, and if properly applied, it makes a very effective finish. Care should be taken to have no wood alcohol in the shellac, as it is likely to cut off the graining color and soften up the paint. Even grain alcohol shellac must be used very expeditiously, and with no small degree of dexterity, or it will cause trouble on the grained surface of the floor. Then again, the graining itself should have been done with special reference to being shellacked instead of varnished. More dryer will be required in the graining color and less oil than on ordinary work.

CHAPTER XXXVII

PATENT GRAINING DEVICES

MACHINES and other devices have been invented for imitating the grains of wood. Many of them are impracticable for ordinary work, and others are effective only when the work can be done on the boards before they are cut up to be made into interior finish.

About 1855 a grainer in London, England, conceived the idea of having the figure of quartered oak engraved on the soft side of sole leather and attaching the leather to a cylinder about ten inches in diameter. The graining color was applied to the work, and the cylinder, made of wood with a metal handle, was pushed over the work, and wherever the leather came in contact with the wet surface, it absorbed the color and left the pattern on the work.

The chief difficulty with this, as with all roller processes, is that the cylinder cannot be gotten into the ends of panels, nor can it be successfully used unless the cylinder is just the size or slightly smaller than the width of the panels.

The principal objection to all mechanical graining is its repetition of pattern, for while the individual piece of work may be excellent, it becomes monotonous when repeated over and over again.

The Mason pad graining machine was in use in 1864, and was made of convex shape similar to an oscillating blotting pad. It was composed of a framework of wood, covered on the convex side with a sheet of plastic compound similar to the material used in a printer's roller. On this surface was engraved the pattern, and it was impressed on the wet graining color by placing one end of the convex pattern at the bottom of a panel, and by rocking the frame, which had handles at either end, the pattern was made in the graining color.

In cold weather the composition on the face of the pad would freeze so hard that it would make no impression on the wet graining color, while in hot weather it would almost melt and run together.

Callow's stencil plates have been in use thirty years or more, and if properly handled, will produce fair to good work. They can be used with greater success by a grainer than by any amateur. The objectionable repetition is their chief fault.

The smooth-faced, large, cylindrical roller, which is

covered with a composition similar to that used on printers' rollers, is one of the best methods of imitating porous woods. The wood to be imitated must be quite porous, and if the pores are not sufficiently deep, they must be eaten deeper into the wood with a strong solution of potash. The board used for the pattern must be perfectly smooth, clean, and dry. Spread the graining color evenly over the board to be imitated, carefully filling all the pores of the wood. Then use a thin piece of wood to scrape off the surplus color, leaving the pores filled. The roller is then passed over the board and picks out sufficient color from the pores to make a well-defined pattern on the roller. This pattern is in turn transferred to the door or other place prepared to receive it by simply rolling the cylinder over it.

Transfer paper of various kinds has been invented for imitating the grains of wood. Some of it is undoubtedly copied from wood, but more of it is imaginary.

The most successful mechanical graining I have ever seen, other than the transfer roller, was the invention of a grainer, William Shannon of Pittsburg, Pennsylvania, who invented a machine for representing the grains of oak, in both heart and quartered grains. His machine chopped the pores into soft wood, having first compressed the grain of the wood on the surface, reducing an inch board to seven-eighths of an inch, then chopping in the pattern with a set of knives, the pores being sunk into the wood one-sixteenth of an inch deep. The machine was made from a second-hand planing machine, and the pores were filled by the same machine while the board was travelling through. The board entered the machine white pine and came through on the other end apparently quartered oak, filled, and ready to be nailed up and shellacked or varnished.

A patent had been granted to W. W. Greer of Hulton, Pennsylvania, for an ingraining machine, which was a cylinder covered with fine teeth which, when rolled over a board, produced imitations of pores in the wood. It was claimed that Mr. Shannon's machine infringed on Mr. Greer's patent, and I believe Mr. Shannon was prevented from doing business with the machine.

Some of the piano manufacturers now have a process of stamping the figure of quartered oak into the grain of rock maple. The figure is made by steel plates with projecting teeth and is sunk deeply into the wood by the use of jack-screws; when these artificial pores are filled and the color of the wood made similar to that of dark oak, it is not easy to detect the deception.

A patent was issued for a belt roller machine which took up the pattern from an etched sheet of plate glass and transferred it to the work by a process similar to that describing the large, smooth-faced roller.

When original pictures are painted entirely by machinery, then and not till then will good hand graining cease to be in demand.

CHAPTER XXXVIII

SHOW PANELS

NO grainer worthy of the name and no young man who aspires to be a grainer should neglect to procure some panels of wood or cardboard and endeavor to faithfully represent both the color and the grain of natural woods, taking for his copy as good examples of the natural woods as he is able to secure.

Much of the idle time of young men in the dull seasons or in the long winter evenings could be put to an excellent use if they would try to improve their work. Sometimes two young men working together will help each other, but individual, patient, painstaking effort is the surest road to success.

If the young grainer is really in love with his business, he will probably have panels done when he first began to work, and they are worth saving; for they will show whether he has corrected his early faults, also the progress he has made, which can be seen by comparison with his panels of later years.

Always have a few panels on hand grounded ready to grain, and then some day when you feel like doing something of a high order or making a copy of some nice piece of wood, you can bring out your panels and begin to work. If you have to first ground the panels, the chances are that something will intervene to prevent you from doing them at once. In these days we can purchase heavily calendered pasteboard with a coated face that readily takes paint and which in some respects is superior to wooden panels, as it will not warp or split and is beautifully smooth. Coat such panels with a rather oily first coat with plenty of dryer or soak them in linseed oil; don't use shellac for a first coat, it makes the cardboard brittle. One more coat of color mixed one-third oil and two-thirds turpentine will cover the panel and prepare it for graining.

It is a good plan to look at your work in a mirror. You can then see how it looks reversed and it may show you chances to improve.

The size of the panels should be about 10 in. × 30 in., or larger if you choose. If made of wood, put a screw eye in the end or make a hole in top of centre of panels so they can be hung up on a nail.

It is a good plan to exchange panels with men in the same line of business in your own or other cities.

A friend of mine, an excellent grainer now deceased, told me about a Grainers' Association of which he was a member many years ago, on the other side of the Atlantic. They held monthly meetings, and at every meeting each man brought a panel grained to represent whatever wood was designated for that meeting. The panels were all the same size and were brought tied in paper, and none but the man who received them knew from whom they came. At the proper time the paper was removed, and each member passed his criticism on the panels, not knowing (unless he was very keen) whose work he was criticising, except his own.

This is an excellent plan and one that might be followed with profit by grainers on this side of the ocean. We are all very likely to adopt certain mannerisms or eccentricities in our work, and intelligent criticism is a healthy thing for us to undergo.

CHAPTER XXXIX

GRAINING ON GLASS

A VERY effective imitation of wood or marble may be done on a smooth piece of glass. Plate glass is the best for this purpose. The work is done on the back of the glass and in just the reverse order from the ordinary way.

The overgraining is first applied, then the graining color, and last of all the ground-color, which backs up the work and brings into view the transparent color already applied to the glass.

A piece of paper or cardboard the size of the panel to be grained should first be prepared with the ground-color. This is placed beneath the glass while the overgraining is being done, and allows the progress of the work to be clearly seen.

It is wise to sponge the glass over with a wash of vinegar before any color is applied. The overgraining may then be done in water color. If the graining is done in oil color, it may, when dry, again be overgrained or shaded in oil color; when the graining color is dry, apply the ground-color.

Excellent imitations of inlaid work may be done by this method, and it is an interesting study when time will permit.

CHAPTER XL

IMITATIONS OF CARVING

THE illustration on the opposite page is a fairly accurate drawing of a ruffed grouse shot by the writer and placed in the position shown by the engraving.

In rooms or halls where the light is rather subdued such work can be successfully done, and it requires considerable skill to produce the proper effect.

There are often places on walls or ceilings where the skill of the grainer can be shown in imitating carved work and mouldings, but it should be carefully done, having due regard for the surroundings, or the effect is disappointing.

IMITATIONS OF MOULDINGS

The grainer is seldom called upon to imitate mouldings, but should he be requested to do so, he should not be found lacking in ability. Considerable technical skill is required to successfully imitate mouldings. A steady hand and a correct eye are very essential.

The beginner should carefully study the light and shade of mouldings whenever he sees them and endeavor to fix in his mind the principles which govern this special line of work. It is well to study the work of some skilful fresco painter and see how he produces the effect of mouldings in light and shade. As a rule the mouldings imitated by the grainer are not as elaborate as those done by the fresco painter.

Mouldings on a grained surface can be laid out by a chalk line or by a lead pencil, and the lines painted with the bristle liner, using a straight edge. It is a good plan in laying out the mouldings to remove nearly all the graining color from two sides of the moulding, leaving the other two sides in the shade. A careful consideration of the situation of the work will determine the proper manner to dispose the light and shade. If the work is seen mostly at night, the source of artificial light cannot well be ignored, but the disposal of the light and shade should be governed by the direction from which the light comes.

Do not paint the colors too strong. Make them harmonize as far as possible with the light and shade on the real mouldings of the adjacent work. After having laid out the mouldings, the lines may be painted in either oil or water colors. The latter method is a good plan if the time is limited, as the work can be at once overgrained in oil and finished. The ground-color with which the work has been prepared will suffice for all high lights on the mouldings, as nothing lighter than the ground-color can possibly be seen on the real mouldings which are grained in the same color.

A short-haired, flat, fresco bristle liner is the best tool for painting the lines of imitation mouldings. The straight edge should not be more than thirty inches in length, and should be bevelled on the back edge so that the color from the brush cannot gather and touch the work. It should be wiped with a rag after every line is drawn to remove any graining color that may have adhered to it.

The grainer who really has at heart the desire to excel in his work must acquaint himself with all the possibilities of his calling, and in his leisure time, when business is dull or in the long winter evenings, he can study the painting of mouldings, using the real articles for his guide, so that when the time comes for him to display his ability on some job, he may be able to attempt it without embarrassment.

Imitations of carving can also be grained in light and shade, and often the effect is as pleasing as though the work were done in color. Care should be taken not to imitate such work where the light is too strong. It is most effective when executed in places where the light is subdued, as it is exceedingly difficult to suggest rotundity on a flat surface in a very strong light.

CHAPTER XLI

CAUSES OF CRACKING IN GRAINED WORK

THE causes of cracking in grained work are a most interesting study. Oftentimes we can only conjecture what has caused the eccentricities displayed by the same varnish in different portions of the same room. Sometimes one panel or stile of a door will crack very badly, while the rest of the door will remain comparatively free from cracks. There is a wide field for investigation, by the student with time at his disposal, to try and discover the causes of these cracks. Many times the causes are quite evident to the skilful workman, while at other times he is absolutely at a loss to even guess these causes.

The excess of liquid dryer in oil graining color is a prolific cause of cracking in the varnish. The dryer, being spread over a non-absorbent groundwork, is left on the surface, and is thus in a position to attack the subsequent coat of varnish.

I have made some experiments to show the result of variously prepared groundworks on the varnish and found that there was little or no difference in the cracks that appear in ordinary varnish over a groundwork prepared of a dead flat color, or one-third oil to two-thirds turpentine, or old fatty color thinned with clear oil. The ground-color was for oak and the graining was done in oil color. All these tests were on the same wide board longitudinally and varnished at the same time. One middle cross-section of the board was left without varnish; the board was exposed to the weather in an easterly direction for six months. At the end of that time fine cracks appeared all over the varnished surface, but, strange to say, the old fatty color groundwork showed cracks no worse than those on the flat groundwork. It is now seven years since this test was made. Since that time the board has been hung indoors, and the cracks appear to have grown no wider or deeper. It would not be well to infer from this test that it makes no difference whether the ground-color is composed of old fatty paint or otherwise. I believe it makes a great difference, notwithstanding the results.

On general principles the groundwork should be hard and non-elastic, or nearly so, if durable work is expected.

After many years of careful observation, in a New England climate, of the effect of the weather on varnish which has been applied to exterior surfaces, I am fully convinced that in a majority of cases such grained work would be more durable without varnish than with it. The reason for this belief can be readily understood when we consider the extremes of temperature to which such work is exposed every year, 100 to 120 degrees, to say nothing of the violent changes of temperature (often 30 to 40 degrees) which take place in twenty-four hours.

The material to which the varnish is attached cannot resist such changes without expanding or contracting, so that if the varnish is of the best quality, it must in time crack, when its elasticity has departed. Therefore, for the past twenty years I have advocated that an outside door grained in oil should be let severely alone for thirty days or more, and then a thin coat of linseed oil, with a few drops of liquid dryer added, will prove a more durable coating than any varnish, because cracks will not appear to destroy the graining and the work can be revived by a coat of linseed oil once a year if necessary.

CHAPTER XLII

THE GRAINER IN FICTION

MANY persons of intelligence have peculiar ideas about graining and the methods by which it is done. But two cases have come to my notice in which reference to grainers is made in the works of writers of fiction. Both are in the writings of the novelist, Charles Reade. In his story, "It's Never Too Late to Mend," he tells us how one Tom Robinson, a character who had been put in jail for stealing, was able by diligent study of the grains of pieces of natural wood (which were supplied him by the kind-hearted chaplain), and being furnished with the proper tools, etc., to become an expert grainer, or "ingrainer," as he calls him. After being sent as a convict to Australia he obtained a ticket of leave and went about graining front doors walnut, oak, mahogany, or satinwood, to the admiration of all beholders. He prepared the ground-color, grained and varnished the door, and got his money all in one day.

Such a rapid execution of the work, while not impossible, would be plainly so if he used oil paints, and the inference is given that such paints were used.

A wager was made by a friend of mine that he could apply two coats of paint to a door, grain, and varnish it in five hours, and he won the wager with hours to spare, doing it in less than three hours. He grained the door mahogany, preparing the groundwork with dry white lead and dry colors thinned with diluted shellac, following one coat of ground-color as soon as the first coat was dry, graining the wood in water colors, and varnishing immediately.

Some of the water-color washable paints are said to make good groundworks for work that is to be rapidly prepared, but as a rule they are not recommended, for reasons given elsewhere.

In a volume by Charles Reade, entitled "Good Stories," one short story is called "Singleheart and Doubleface" and among other characters is one James Mansell, a painter and grainer, who was the successful suitor for the heroine of the story. Mr. Reade writes "Mansell had three trades. In one of them (graining) he might be called an artist. He could imitate the common woods better than almost anybody, but at satinwood, mahogany, and American birch, he was really wonderful."

After marrying the heroine and settling down he acquired the habit of indulgence in intoxicating liquors to such an extent that he lost both his self-respect and his customers. Mr. Reade writes, "Mansell was styled the first grainer in the place and the tradesmen would have employed him by preference if he could have been relied on to finish his jobs, but he was so uncertain; he would go to dinner and stop at a public-house, would appoint an hour to commence, and be at a public-house." "He tired out one good customer after another, the joint income declined in consequence, and, as generally happens, their expenses increased, for Mrs. Mansell getting no help from her husband was obliged to take a servant."

"Often in the evening she would close her shop early, leave her child under strict charge of the girl, and go to some public-house and there coax and remonstrate, and get him away at last."

* * * * * * * * *

"At last it came to this, that nobody in the town who knew James Mansell would employ him."

* * * * * * * * *

"This man's vanity was prodigious; it equalled his demerit."

Mansell finally died a drunkard in America.

Here is faithfully portrayed for us, by a master hand in fiction, the end that often follows the course of the grainer who allows himself to be led away by his appetite for strong drink.

It is the curse of many a skilful workman and particularly in the graining trade. Many such I have known, and their work I have admired, but in no case was strong drink any help to such men. On the contrary, it was their ruin. No man ever did, or in my opinion ever will, excel in whatever sphere of usefulness his work lies while he is in an artificial condition. The man who totally abstains from intoxicating drinks, while he may not have the natural talent that is the gift of many, yet his work, if conscientiously and faithfully done, will often surpass the work of the man who is brilliant and mediocre by fits and starts, according to the condition of his mind when doing his work. It is often possible, by careful examination, for an expert grainer to tell with tolerable accuracy the physical and mental condition of the man who did a piece of work.

A word of advice to all young men who aspire to be grainers: abstain from intoxicating drinks and I warrant your work will be improved thereby. Nor is the use of tobacco in any way essential to your success. A man's mind ought to be entirely free to allow him to concentrate all his efforts to the successful accomplishment of the work in hand, and making a chimney of the mouth is in my opinion "a wasteful and ridiculous excess." It is neither eating nor drinking. It takes time and money from the workingman, and unless prescribed by a physician, for some special reason, I can see no use for it. I have heard an intelligent master painter, himself an inveterate smoker, say that the man who did not use tobacco was worth ten cents a day more than the one who used it, for he could devote all his time to the work and was not disturbed by seeing others smoking or chewing tobacco.

CHAPTER XLIII

GRAINING A DOOR QUARTERED OAK

An Illustrated Talk given at the Second Annual Convention of the Master House Painters and Decorators Association of Canada held at Hamilton, Ontario, July 25, 26, 27, 1905.

MR. PRESIDENT AND GENTLEMEN: —

I don't wish to present myself to you to-day as anything else than an humble imitator of nature; but if any of the little things shown are new to you, it may be of some help in getting closer to nature in your work.

First, we should have a proper foundation of ground-color, which should be strained through fine cloth before being applied and thinned with about one-fourth raw linseed oil to three-fourths spirits of turpentine for old work, adding a larger proportion of linseed oil for new work. A little varnish may be added to the last coat, as it tends to hold the color and gives a better surface to work on. A sufficient quantity of dryer is also added. When the last coat is dry, it should be lightly sandpapered, and then we are ready to apply the graining color. For representing quartered oak this color is generally mixed in oil, although good work can be done in water colors, but by a different process. In case a fair to good job is wanted, and we cannot spare time to come back and overgrain the work, we may first apply a thin glaze of water color directly to the groundwork. I will treat this panel in this way, leaving the others plain.

For the glazing color we use a little diluted drop-black, ground in water and thinned with one part stale beer and one part water. If the color creeps or crawls or will not readily attach itself to the ground-color, we will use some bolted whiting, which, on being rubbed over the panel, will effectually stop the cissing or crawling. If the oil color crawls, the same treatment may be given, or a better plan is to first dampen the work with benzine. This is the most effective process to prevent the crawling of any kind of paint or varnish and it in no way effects the durability of the work.

We now take a short-haired overgrainer, and after wetting it and charging it with the thin color we separate the hair with an ordinary bone comb and apply the color to the panel. Before the color has time to dry use a steel comb to serrate the regularity of the lines.

We will now mix our oil graining color. Supposing we want to have the work of a medium shade, the color can be mixed about one-third burnt umber to two-thirds raw sienna, adding a little drop-black to subdue the brightness of the other colors. This color we will thin with a mixture of two-fifths raw oil to three-fifths turpentine, using about one-half pint of good liquid dryer to the gallon of mixture. In this mixture we dissolve about two-thirds of an ounce of yellow beeswax, which we first cut in shavings and melt in an iron vessel, adding turpentine slowly after taking it from the fire, or the wax may be cut in shavings placed in a wide-mouthed glass bottle, and the bottle filled two-thirds of its height with turpentine. If this be done overnight and kept in a warm room, the turpentine will have so softened the wax by morning, that a few violent shakings of the bottle will finish its dissolution. It can then be added to the gallon of thinners. For many reasons I prefer the finest grade of dry colors, except black and burnt sienna, in preparing graining color.

We now work in our flat rubbing-in brush, and apply the color in the usual way. When it has slightly set, we comb it with a rubber comb and intersect the tracks of the rubber comb with the steel comb. This gives the porous appearance of the wood, and we imitate the quartered veins by wiping off the color with the rag drawn over the end of the thumb nail. We can blend the edges of the work by using the second joint of the forefinger, or we can use a short-haired fitch for this purpose. We can blend the work lightly with a rubbing-in brush, always in one direction.

A piece of straw matting makes an excellent fine comb for quartered oak. To get a darker effect for some of the veins we take some of the rubbing-in color, and with a flat fresco lining fitch apply the color directly to the combed work and blend it quickly with the rubbing-in brush. The effect produced is not as good an imitation of wood as if the work is done on dry color, but it gives variety to the work, and if carefully done, is a fair representation of the darker veins of quartered oak.

In graining the cross stiles of the door we must take pains to go well beyond the joints, and if the long stiles are to be done light, we cover the coarse or medium-toothed steel comb with one thickness of the rag and draw the mitre lines cleanly, combing the rest of the stiles with the same comb, or if necessary using the rubber comb outside the mitre joints. We then overcomb the tracks of the rubber comb with a finer steel comb. The work of wiping out the veins or putting them in with a fitch can then be done.

If the long stiles are to appear as slightly darker than the cross stiles, use a sash tool to cut the mitres, having a dip of color from the bottom of the pot. The most effective process is to do the long stiles of the same color as the rest of the work, and when dry, overgrain them to the depth of color desired, or they can first be glazed with a thin wash of water color as previously described.

If the doors in a room are nearly or quite dry before we leave the job, we can overgrain the more prominent portions of the figured work by using the short-haired overgrainer in the oil color, separating the bristles with a comb. Take care to have the color quite thin, and the steel comb can be used as directed in the use of the overgrainer in water color. A good imitation of any kind of wood is rarely done without being overgrained. Some kinds of woods require more attention in this respect than others, but all can be helped by overgraining. Two or three days should elapse before the work should be varnished.

Gentlemen, I have briefly tried to show you how a fair imitation of quartered oak may be done, and I trust you may have gained something in the way of information from what you have seen and heard.

CHAPTER XLIV

NEW METHODS

IN the development of new methods a fertile field is available for study to those who care to depart from conventional lines of working. The patient investigator will find that the field has been well worked over by his predecessors. Still, he will find ample opportunity to improve his work, and possibly his method of working, by a diligent study of the wood as he finds it in his particular locality.

The grain of many woods is naturally affected by the soil in which it grows, or by its geographical location. An oak tree grown in Canada may have very different markings in its grain from that of a similar species of tree grown in a southern clime and on a different soil. The layers of wood in trees grown in temperate climes are, as a rule, produced in annular rings; yet a friend of mine, a botanist, and one not likely to be mistaken, informed me that he had seen pine trees growing in the rich alluvial bottom lands of Arkansas that added three tops to their growth annually. This meant three rings to the growth of the tree, so that a grove of trees whose age he estimated to be thirty years were in reality planted but ten years. They had been set out by the man in charge of the estate, and my friend told me that he would not have believed that the trees were so young had he not seen the marvellous rapidity with which they grew.

Some varieties of quartered oak may require an entirely different method of treatment from that ordinarily used in order to make a successful imitation. It is a wise plan to endeavor to be versatile and not accustom one's self to a particular method of working. Be governed by certain principles rather than by rules. One man may fail to achieve success by the process which he is taught, and yet he may accomplish excellent results by another process, possibly by one he may himself have developed. In any case the persistent study of the grains of wood is essential to any person who desires to become an expert imitator. Examine the same piece of wood in a strong sunlight, also by artificial light. Notice the play of light and shade that often travels across it as you move the board or change your point of view.

The grainer cannot produce (except to a very limited extent) this iridescence, but he should know as much as possible about it and endeavor to adjust his work in conformity to the markings he observes in the wood. These shadings are often represented much stronger and darker than they appear in the wood, but this is a great mistake.

One of the oldest methods for producing an iridescent effect is to have a metallic leaf groundwork. Gold or silver leaf is usually employed. Aluminum leaf or bronze is more frequently used in these later days. It is better to leave the work as grained, or, if done in water color, to oil it over and wipe off any surplus oil with a soft cloth, as varnish largely destroys the effect.

In imitating light to medium, or even dark, quartered oak, it is doubtful if a more effective process can be employed to produce the iridescence than to use for a foundation for the work some light, clean-grained wood, such as clear pine, spruce, or whitewood. It may be prepared by a thin coating of white glue-size or, better, by two thin coats of white shellac. When dry, apply the graining color and grain in either water or oil color, varnishing as usual; or it may be finished in white shellac, which is probably the better plan, as it avoids the possibility of the cracking of the varnish due to lack of affinity with the shellac under coats. Care must be taken to use as little linseed oil as possible in the graining color.

The work may also be prepared by coatings of liquid filler or pale oil varnish. The former often contains elements which utterly destroy not only the graining, but all subsequent coats of varnish, hence its use is not recommended. The oil in the varnish is apt to discolor the new wood more than white shellac. This is a factor not to be overlooked where it is desired to keep the work as light as possible. It is very seldom that this process can be successfully applied to an entire room, as the wood is rarely sufficiently free from imperfections to allow its use over large surfaces. It is better adapted to panels or comparatively small surfaces, but when carefully done, it surpasses in woodiness and **transparency** work done on any solidly painted groundwork.

In passing it may be worth while to notice that the mottled effect of nearly all woods, and the shadows that appear in connection with the heart grains of nearly all woods, are due to the angle at which the fibre of the grain approaches the surface of the wood. When the fibre runs parallel, the grain is apt to appear without mottled effect. The open ends of the fibre present a dark appearance, and when the undulations in the tree are intersected by the saw, an effect is produced more or less mottled, according to the character of the wood.

CHAPTER XLV

JOURNEYMEN

IN the early days of the trade guilds, centuries ago, a workman was known and recognized by his ability rather than by the quantity of work he was able to accomplish in a given time. After faithfully serving his apprenticeship, generally for a term of seven years, he was still not recognized as a journeyman until he had, as the name implies, made a journey of many miles and worked at his trade in several cities. On his return home he was recognized as a full-fledged journeyman and considered competent to be admitted to fellowship with his fellow-workers. In his travels he was provided by the guild with credentials certifying that he had duly served his apprenticeship. Wherever a workman in his trade was needed, he was given employment, and the local guild looked after his welfare. This system broadened the ideas of the young craftsman and allowed opportunity for comparison to be made between the work and methods of his late master and those of skilful craftsmen of other cities and towns. It also put him on his mettle and stimulated him to do the best and prove his claims to be an expert workman.

These facts are cited to draw attention to the great benefit to be derived by the young grainer in studying work and methods of skilled grainers in cities or towns in their vicinity or in any city in which they may find themselves; the object being not to copy the work of any man, however excellent in itself, but to study the process by which results are attained, to avoid any fault observed, and, if possible, to improve on the work examined — always in the direction of a closer imitation of nature.

To the expert workman, in his examination of a piece of finished work, it need not be necessary to explain the process by which it was done. Oftentimes he can determine with tolerable accuracy the methods employed and tools used, also the number of stages through which the work passed before it was completed. It is true that even the most expert grainer may be mistaken in his opinion of how the work was done, but this is seldom the case. Even the bank clerk is sometimes mistaken and allows a counterfeit bill to pass his scrutiny without detection.

It is often a revelation to the aspiring young craftsman to observe the results obtained by a workman of another school from that in which he was taught. Tools and vehicles for applying the color may be radically different from those he may have considered as standards. It is a wise man who can profit by the mistakes of others or who will try and correct his own shortcomings. It is remarkable how a simple process may be used to produce effects which to the beginner seem marvellous. He should, as far as possible, acquaint himself with all known processes by which wood can be imitated and endeavor to master them so that his work shall not appear as a mere copy of the work of his master, but shall at least suggest the grains of the wood which he is trying to imitate.

CHAPTER XLVI

BICYCLE FOR COUNTRY OR CITY WORK

THE grainer who works for the trade outside the very larger cities is losing an opportunity to earn extra dollars if he fails to use a bicycle. There are many occasions when he can get home to dinner and save both time and money by using a bicycle. A grainer's kit is so small that it is not difficult to carry it on a wheel, and a little practice will soon give one confidence so that he can safely go from place to place and save hours of valuable time.

Possibly the grainer of the future may ride to and from his work in an automobile, and his "rubber in" may act in the dual capacity of rubber in and chauffeur, but the prices now paid for work will have to be raised one hundred per cent before such things are likely to transpire.

NOTE. This prediction has come true. The trade grainer now rides to work in an automobile, his "rubber-in" acting as chauffeur, and prices have advanced over one hundred per cent in ten years.

CHAPTER XLVII

BUTTERNUT

Ground-color. — White lead, raw sienna, raw umber.

Graining Color. — Raw sienna, raw umber, burnt umber.

Tools for Oil Color. — Rubbing-in brush, sash tool, fitch tools, bristle liner, combs, rags, crayons.

Tools for Water Color. — Sponge, rubbing-in brush, stippler, blender, sash tool, fitch tools, bristle liner, crayons, overgrainers.

This wood is found in the middle western states and is sometimes called white walnut. Its grains are quite similar to those of black walnut, but are as a rule more angular in the heart grains, and the outline of the grains is much less vigorous. There is but little mottle to its grain, and it has a quiet and subdued effect.

Some of its grains are not unlike those of white mahogany. It is seldom seen in the East nowadays.

For oil color mix about one-third each of raw sienna, raw umber, and burnt umber. Thin with the regular thinners, making the color very thin.

The work can then be rubbed in, and the faint stippled effect can be produced with the dry rubbing-in brush on the color which has been allowed to set slightly.

The heart grains can then be put in with the flat fresco liner and the work blended with the rubbing-in brush, care being taken not to lift the color too much in blending.

Combs can be used for the finer grains, and the sides of the rubbing-in brush can be used to remove portions of the color. The stippling with the dry brush can then be done over the combing.

For water color graining the work is first sponged over, using one-third beer to two-thirds water and rubbed in with a mixture of raw sienna, raw and burnt umber, using the rubbing-in brush, and applying the color very sparingly. Stipple at once, and when dry, the heart grains may be put in with the bristle liner, and blended quickly, or the dry crayon may be used for this purpose.

The plain grains can be represented by using the short-haired overgrainer.

Remember that the outline of the heart grains should not be too pronounced. The crayon outline is often nearer to the natural grains than is the work done by the brush.

CHAPTER XLVIII

GUMWOOD

Ground-color. — White lead, raw umber, raw sienna.
Graining Color. — Raw umber, drop black.
Tools. — Sponge, rubbing-in brush, blender, sash tool, fitches, overgrainers, medium steel combs, straw matting, soft cotton rags.

This wood is a native of the southern parts of the United States and in recent years has come into fashion for interior finish and for furniture. Much of the furniture now sold for walnut is really gumwood with a thin walnut veneer on the more important surfaces.

The grains can be represented in either water ($\frac{1}{2}$ vinegar) or oil colors. The figures are slightly darker than the background of the wood and often interlock in a most peculiar manner. These grains are generally longitudinal and seldom have any mottled appearance. In some respects they resemble the figures of rosewood.

In working from a water-color base, a small portion only can be done at one time. Rub in the color (composed almost wholly of raw umber) and with a sponge clean off or lighten up longitudinal streaks and blend softly, then with color darkened slightly with a touch of drop black introduce the figures with the fitch tools and blend again before the color dries. This process can be used again after the color is dry, but do not overdo the work nor make too strong contrasts.

In working in oil colors, make the raw umber very thin and with a soft rag remove portions of the color and blend softly with the rubbing-in brush. If necessary, use the steel combs (always covered with thin cotton cloth) or the straw matting for the plainer portions of the work; then, having darkened the color with a very little drop black, put in the darker veins with the fitch tool or bristle fresco liner.

It is sometimes necessary to add a little dry zinc to the oil color to produce the gray shades seen in the wood. Bolted whiting helps to produce this effect.

CHAPTER XLIX

DOUGLAS OR OREGON FIR

Ground-color. — White lead, yellow ochre, raw sienna.
Graining Color. — Raw sienna, burnt sienna, raw umber.
Tools. — Rubbing-in brush, sash tool, fitch tools, rubber and steel combs, soft cotton rags.

This wood is now much in evidence for doors and standing finish in modern houses. It grows in the northwestern portions of the United States and is a sound and durable timber. Much of it is used in a veneer form, being cut from the log around the outer circumference, and in this way it presents a wonderful diversity of grain, always continuous, and sometimes of gigantic size; and again it will run very fine in grain.

It can be represented very faithfully in oil color. Mix raw sienna and raw umber and thin to a very light stain; rub in cleanly and rather sparingly, then with a soft piece of cotton rag remove portions of the color in the general directions of the grain of the wood. Add a little burnt sienna and a touch of raw umber to the graining color (keep this in a separate vessel), and try to imitate the figures produced by the peculiar way the veneer is cut. When partly set, blend lightly with the rubbing-in brush, and if necessary use the combs to carry out the lines on the outer edges of the work.

The plain grains can be made with the rubber or steel combs (the latter always covered with a rag), and blended lightly, always lengthwise of the grain.

INDEX

[Roman figures pages, black figures plates.]

A

Antiquity of graining, 1.
Ash, 46; burl, 53; ground-colors for, 46; Hungarian, 51; wiped out and pencilled, 48.

B

Badger blender, **1**.
Bicycle, for city or country work, 129.
Bird's-eye maple, 35; maple in water color, 36.
Black walnut, 87.
Blending groundwork, 12.
Bristle rubbing-in brushes, **2**; liners, fitches, **2**; mottlers, overgrainers, **1**; stippler, **1**; piped overgrainer, **1**.
Burl ash, 53; ash, ground-colors for, 46; ash in oil color, 54.
Burl walnut, 91.
Butternut, 129.

C

Camel's-hair piped overgrainer, **1**.
Case for carrying tools, 131.
Causes of cracking in grained work, 116.
Ceilings, 102.
Champs, to put in (quartered oak), 57.
Check roller, 21, 30.
Cherry, 81; groundwork for, 81; in oil color, 82; in water color, 83; mottled, to overgrain, 83; mottled and overgrained, 84.
Chestnut, 71.
Circassian walnut, 93.

Coats, thin preferable, **13**.
Color, for graining, **13**; straining before thinning, 13.
Combing in oil color, 25; combing in water color, 29; combing a background for quartered oak, 27, 28.
Combs, **2**; bone, **1**; cork, 21; rubber, 20; steel, 20.
Covering teeth of combs, 26, 27.
Crayons for light oak and ash, 31.
Curly birch, 85; curly maple, 33; curly walnut, 91.
Cypress, 77.

D

Dark oak, 54; ground-color for, 54; veins in mahogany, 95; veins in quartered oak, 59–62; veins in rosewood, 100.
"Docked" pencil in bird's-eye maple, 37.
Dry colors for graining, 16.
Dryers in graining color, **13**.

E

Eminent grainers of the last century, 7.
English oak, 65.
Eyes and shadows in bird's-eye maple, 40.

F

Feathered mahogany, 95, 96.
Fitch tool, 21.
Flat brush for rubbing in, **2**; fresco bristle liner, **2**.
Floors, 104.

INDEX

Fourteen ways of imitating quartered oak, 60.
French walnut burl, 91.

G

Grainer, in fiction, the, 118.
Grainers' combs, 20, 21; tools, 20, 21.
Graining, antiquity of, 1; both sides of the same panel, 131; colors, 16; crayons, 31; door, quartered oak, 121; on glass, 113; over old paint, 14; quartered oak, 58.
Ground-colors, 11; for ash, 46; for burl ash, 46; for Hungarian ash, 51; for birch (curly), 85; for cedar, 72; for cherry, 81; for chestnut, 71; for cypress, 77; for mahogany, 94; for mahogany (white), 45; for maple, 35; bird's-eye maple, 35; curly maple, 33; silver maple, 40; for oak, 54; English oak, 65; pollard oak, 66; root oak, 67; quartered oak, dark, 54; quartered oak, light, 54; for pine (hard or pitch), 75; for pine (yellow), 74; for rosewood, 99; for satinwood, 42; for sycamore (quartered), 78; for walnut, 87; black walnut, 87; Circassian walnut, 93; curly walnut, 91; French walnut (burl), 91; Italian walnut, 93; for whitewood, 41; for butternut, 129.

H

Hard pine, 75.
Heart grains, in ash, 77; in cherry, 84; in oak, 59; in walnut, 88; in hard pine, 76; in yellow pine, 75.
Heart, or sap, oak, 67.
High lights in bird's-eye maple, 38.
Hungarian ash, 51.

I

Imitations, 4; of carving, 114; of mouldings, 114.
Italian walnut, 93.

J

Journeymen, 127.

M

Mahogany, 94; mahogany (white), 95.
Manila paper for covering a poor floor, 106.
Maple, 35; bird's-eye, 35; curly, 33; silver, 40.
Mason pad machine process, 109.
Megilp, for oil colors, 18; for water colors, 19.
Mixing, graining color, 16; ground-colors, 12.
Mottlers, 1.

N

New methods, 124.
Notes, 132.

O

Oak, 54; English, 65; graining in water color, 67; pollard, 66; root, 67; quartered, 54.
Old paint, graining over, 13, 14.
Old varnish, to remove, 14.
Overgrainers, plain, 1; plain bristle, 1; piped bristle, 1.

P

Paint removers, 14.
Patent graining devices, 108.
Piped bristle overgrainer, 1.
Pitch pine, or hard pine, 75.
Pollard oak, 66.
Preparing old work for graining, 14; ground-colors, 12.
Primary coats, 14.

Q

Quantity of megilp, 18.
Quartered oak, 54; sycamore, 78.

INDEX

R

Removing old paint, 14; **varnish**, 14; graining color from the hands, 132.
Roller, the check, 30; rubber, 109.
Root of oak, 67.
Rosewood, 99.
Rubber combs, 20.
Rubbing, in graining, 21; in oil color, 21, in water color, 24.

S

Sandpapering, 15.
Satinwood, 42.
Shading quartered oak, 57.
Show panels, 111.
Silver maple, 40.
Solution for removing figure of oak, 64; to remove varnish or paint, 14.
Steel combs, 2.
Stippling for mahogany, 88; for walnut, 88.
Sunday work, 131.
Sycamore, quartered, 78.

T

Teak, 98.
Thinners, for oil colors, 17; for water colors, 18.

Thumb nail, substitutes for, in graining, 64.
Tools, for putting bird's-eyes in maple, 37; used by grainers, 20.
Touching quartered oak with solution, 64.
Transfer, paper, 110; roller, 110.

V

Varnishing a grained floor, 108; overgrained work, 101.
Vinegar, use after removing old paint, 14.

W

Walnut, black, 87; Circassian, 93; curly, 91; French burl, 91; Italian, 93.
White mahogany, 45; Oregon cedar, 72.
Whitewood, 41.
Wiping out, in oil color, 57; heart grains in oil color, 32; heart grains of oak, 69.
Wooden wedge, to hold door open, 132.

Y

Yellow pine, 74.
Gum wood, 131.
Oregon fir, 132.